21世纪高等教育计算机规划教材

Java编程技术与项目实战

（第2版）

JAVA PROGRAMMING TECHNOLOGY AND PROJECT TRAINING

(2nd edition)

王诚 梅霆 李琴 王峰 朱书眉 ◆ 编著

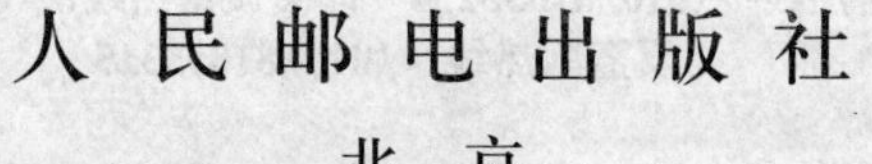
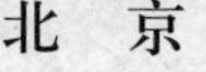

人民邮电出版社

北京

图书在版编目（CIP）数据

Java编程技术与项目实战 / 王诚等编著. -- 2版
. -- 北京 : 人民邮电出版社, 2015.12
21世纪高等教育计算机规划教材
ISBN 978-7-115-40095-6

Ⅰ. ①J… Ⅱ. ①王… Ⅲ. ①JAVA语言－程序设计－高等学校－教材 Ⅳ. ①TP312

中国版本图书馆CIP数据核字(2015)第174140号

内 容 提 要

本书介绍了 Java 语言编程基础及 Java Web 开发的基本知识，主要内容包括 Java 语言编程基础、面向对象编程基础、线程和网络编程、Java 图形用户界面、Java 数据库编程、JSP、Servlet、JavaBean 编程基础、MVC 模式及 Struts2 编程框架，最后给出两个具体实例。本书基本覆盖了 Java 编程及 Java Web 编程的大部分实用技术。本书编排结合具体真实项目实例展开教学，由浅入深，重点突出。

本书在选择编排内容时注重实用性，强调实践技能的提高，同时也注重系统性、科学性和学习性，每章末都有练习。本书层次分明，理论联系实际，用典型的案例来演示 Java 编程技术的魅力，有利于培养工程应用及编程能力。本书既可作为高等学校计算机软件及应用、电子信息工程、通信工程、软件工程等相关专业的本科生教材，也可供广大 Java 软件开发工程师学习参考。

◆ 编　　著 王　诚 梅　霆 李　琴 王　峰 朱书眉
责任编辑 武恩玉
责任印制 沈　蓉 彭志环

◆ 人民邮电出版社出版发行　　北京市丰台区成寿寺路 11 号
邮编 100164　　电子邮件 315@ptpress.com.cn
网址 http://www.ptpress.com.cn
北京昌平百善印刷厂印刷

◆ 开本：787×1092 1/16
印张：16.5　　2015 年 12 月第 2 版
字数：435 千字　　2015 年 12 月北京第 1 次印刷

定价：39.00 元

读者服务热线：(010)81055256　印装质量热线：(010)81055316
反盗版热线：(010)81055315

第 2 版前言

当前，Internet 正以惊人的速度发展，计算机技术、软件技术、物联网技术、网络技术、通信技术等 IT 领域新技术不断涌现，使现在的生活出现了革命性的变化，特别是物联网技术的兴起，推动了各行业和部门的智能化、智慧化应用研究。在这些信息技术的应用中，Java 技术凭借其独有的、与平台无关的、与网络紧密结合的特点，在企业级应用及移动终端的应用编程中发挥越来越重要的作用。

本书在介绍 Java 基本知识的基础上，侧重介绍 Java Web 的框架知识，并给出项目的原型实例。本书从最基本的下载配置和安装开始教学，从最基础的数据类型、表达式、语句讲起，引导读者进入面向对象的编程环节中；中间贯穿了线程技术、网络技术、数据库技术、Java Web 的基本框架等方面的编程知识；最后结合两个具体的实例来进行应用编程的总结。

本书并不要求读者逐章阅读，每一章内容都可独立学习，读者可以根据应用需要进行选择。本书的实例都是可以运行的，建议读者在学习过程中录入书中的例子以加强实践，以便更好、更快地掌握技术要领。

本书的先修课程包括 C/C++语言程序设计、计算机编程基础、网络技术、数据库技术等。如果读者学习并掌握 C/C++语言的使用，则学习本书可能会轻松一些，因为它们的语法结构是相似的。如果读者具备面向对象编程的基础，则更容易掌握 Java 编程技术。

本书在编写过程中得到了很多同行和读者的帮助，在此感谢南京邮电大学的陈杰、许晓、范向阳在整理资料和校对等方面所做的大量工作，以及宋文广、赵振文工程师提供的项目代码实例，在此一并表示感谢！

由于时间仓促和编者的知识所限，书中难免存在错误与不足之处，欢迎读者批评指正。作者联系方式为：wangc@njupt.edu.cn。

编　者

2015 年 5 月

目录

第1章 Java语言概述及编程基础

Java语言是Sun公司（已被Oracle公司收购）于1995年正式推出的面向对象（Object-oriented）的程序设计语言。Java语言由于具有安全、跨平台、面向对象、简洁等特点，一经推出即引起了广大软件公司和程序员关注，受到计算机界的普遍接受与欢迎，成为目前网络时代最为流行的程序设计语言。Java问世以来，其技术发展非常快，在计算机、移动电话、家用电器等领域都得到了广泛的应用。

本章将简单介绍Java语言的概述、Java环境的安装与配置，以及基本的Java语法知识。通过本章的学习，同学们可以对Java语言的发展、特点和语法知识有基础的理解，并且可以在自己的工作机上安装搭建Java开发环境，为后续的深入学习做准备。

1.1 Java语言概述

1.1.1 Java语言发展

1991年4月，Sun公司的James Gosling领导的绿色计划（Green Project）开始着力发展一种分布式系统结构，使其能够在各种消费性电子产品上运行。他们研发了一种新的语言，该语言以C和C++为基础，James 根据他在Sun公司办公室外的一棵橡树，而称其为Oak语言。后来发现已有一种称为Oak的计算机语言。当一些Sun公司的员工到当地一家咖啡店时，有人提议将该语言命名为Java，从而使Java这个名字一直延续至今。

不过，在当时市场不成熟的情况下，他们的项目没有获得成功。市场对智能型电子装置需求的上升率并不像Sun司所期盼的那样快，更糟的是Sun公司参加竞争的一个重要的销售合同被另一公司得去了。此时，Green项目几乎处于被取消的境地。但很幸运的是，1993年万维网（WWW）疯狂地流行起来，由于Internet的迅猛发展和WWW的快速增长，第一个全球信息网络浏览器Mosaic诞生了。此时，工业界非常急迫地需求一种适合在网络异构环境下使用的语言。Games Gosling 决定改变绿色计划的发展方向，对Oak进行了小规模的改造。就这样给该项目重新注入了生机，1995年，Oak语言更名为Java语言。Java的诞生标志着互联网时代的开始。

Sun公司于1995年5月在一个重要会议上正式发布了Java。这样的事通常不会引起广泛的注意，但是由于万维网的商业利益，Java立即引起了商业界的极大兴趣。目前，Java被广泛应用于创建具有动态的、交互内容的Web网页，开发大规模企业应用程序，增强万维网服务的功能，向消费者的设备提供应用程序。

1998 年是 Java 迅猛发展的一年。在 1998 年 12 月 4 日，Sun 发布了 Java 历史上最重要的一个版本：JDK 1.2。这个版本的发布标志着 Java 已经进入 Java 2 时代。这个时期也是 Java 飞速发展的时期。在这一年中 Sun 发布了 JSP/Servlet、EJB 规范，以及将 Java 分成了 J2EE、J2SE 和 J2ME。

2000 年 5 月 8 日，Sun 对 JDK 1.2 进行了重大升级，推出了 JDK 1.3。2002 年 2 月 13 日，Sun 公司发布了 JDK 历史上最为成熟的 JDK 1.4 版本。2004 年 10 月，Sun 发布了期待已久的 JDK 1.5 版本，同时将 JDK 1.5 改名为 J2SE 5.0。最新版本 JDK 8.0 加入了很多新特性，是一款革命性开发平台，全面升级了现有的 Java 流程模式。

Java 问世以来，其技术发展非常快，在计算机、移动电话、家用电器等领域都得到了广泛的应用。

1.1.2 Java 语言的特点

Java 语言的前身是在 C++的基础上开发的，它继承了 C、C++语言的优点，增加了一些实用的功能，使 Java 语言更加精炼。Java 摒弃了 C、C++语言的缺点，去掉了 C、C++语言中的指针运算、结构体定义、手工释放内存等容易引起错误的功能和特征，增强了安全性，也使其更容易被接受和学习。Java 是独立于平台，面向 Internet 的分布式编程语言。

Java 是一种简单的、面向对象的、分布式的、解释执行的、健壮的、安全的、结构中立的、可移植的、高效率的、多线程的和动态的语言。

1. 简单

Java 是一种纯面向对象的语言，它通过提供最基本的方法来完成指定的任务，只需理解一些基本的概念，就可以用它编写出适合于各种情况的应用程序。Java 略去了指针、运算符重载、多重继承等内容，并且通过实现无用信息自动回收，大大简化了程序设计者的内存管理工作。同时，Java 很小，基本的解释器及类支持大约仅为 40KB。

2. 面向对象

Java 是一种纯面向对象的语言，Java 的核心是面向对象编程。事实上，所有的 Java 程序都是面向对象的，这一点与 C++语言不同，因为在那里可以选择是否面向对象编程。Java 程序面向对象的设计思路不同于 C 语言基于过程的程序设计思路。面向对象程序设计，具备更好地模拟现实世界环境的能力和可重用性。它将待解决的现实问题转换成一组分离的程序对象，这些对象彼此之间可以进行交互。一个对象包含了对应实体应有的信息，以及访问和改变这些信息的方法，重点放在数据（即对象）和对象的接口上。通过这种设计方式，所设计出来的程序更易于改进、扩展、维护和重用。Java 语言只支持类之间的单继承，但支持接口之间的多继承，并支持类与接口间的实现机制。

3. 分布式

Java 是一种分布式的语言。传统的基于 C/S（客户端/服务器）架构的程序，客户端向服务器提出服务请求，服务器将程序执行结果返回，所以，服务器负荷较重。Java 采用 Java 虚拟机架构，可将许多工作直接交由终端处理，因此，数据可以被分布式处理。此外，Java 类库的运用，大大减轻了网络传输的负荷。Java 类库包含了支持 HTTP、FTP 等基于 TCP/IP 的子库。Java 应用程序可凭借 URL 打开并访问网络上的对象，其访问方式与访问本地文件系统几乎完全相同。Java 的网络功能强大且易于使用，特别是远程方法调用使得分布式对象之间可以互相通信。

4. 高效解释执行

Java 是高效解释执行的语言。高级语言程序必须转换为机器语言程序才能在计算机上执行。Java 程序在编译时并不直接编译成特定的机器语言程序，而是编译成与系统无关的“字节码

（bytecode）”，由 Java 虚拟机（Java Virtual Machine，JVM）来执行。JVM 使得 Java 程序可以“一次编译，随处运行”。任何系统只有安装了 Java 虚拟机后，才可以执行 Java 程序。JVM 能直接在任何机器上执行，为字节码提供运行环境。

5. 健壮性

Java 是健壮的语言。Java 不需要指针就可以构造其他语言中需要指针构造的数据结构，即不会存取“坏的”指针而造成内存分配、内存泄露等错误。在传统的编程环境下，内存管理是一项困难、乏味的工作。例如，在 C 或 C++语言中，必须手工分配、释放所有的动态内存。如果忘记释放原来分配的内存，或是释放了其他程序正在使用的内存时，系统就会出错。同时，在传统的编程环境下，对异常情况必须用既繁琐又难理解的一大堆指令来进行处理。Java 通过自行管理内存分配和释放的方法，从根本上消除了有关内存的问题。Java 提供垃圾收集器，可自动收集闲置对象占用的内存。Java 提供面向对象的异常处理机制来解决异常处理的问题。

6. 安全

Java 是安全的网络编程语言，使用 Java 可以构建防病毒、防篡改的系统。Java 提供了一系列的安全机制以防恶意代码攻击，确保系统安全，如禁止运行时堆栈溢出、禁止在自己处理空间外破坏内存。Java 的安全机制分为多级，包括 Java 语言本身的安全性设计，以及严格的编译检查、运行检查和网络接口级的安全检查。

7. 结构中立

Java 是结构中立的语言。Java 的设计目标是要支持网络应用。Java 编译器会产生一种具备结构中立性的对象文件格式，即 Java 字节码文件。精心设计的 Java 字节码不仅很容易在任何机器上解释执行，还可以快速翻译成本机的代码。

8. 可移植性

结构中立是确保程序可移植的必要条件，此外，还需要很多其他条件的配合。Java 在可移植性方面做了许多工作。Java 通过定义独立于平台的基本数据类型及其运算，使数据得以在任何硬件平台上保持一致。例如，Java 中的 int 类型永远是 32 位的整数，这样就消除了代码移植时的主要问题。

9. 高效率

Java 是高效率的语言。每一次的版本更新，Java 在性能上均做出了改进。在历经数个版本变更后，Java 已经拥有与 C/C++同样甚至更好的运行性能。当 JVM 解释执行 Java 程序时，Java 实时编译器（Just-In-Time，JIT）会将字节码译成目标平台对应的机器语言的指令代码，并将结果进行缓存，因为 JVM 能够直接使用 JIT 编译技术将经过精心设计的字节码转换成高性能的本机代码，所示 Java 可以在非常低档的 CPU 上顺畅运行。

10. 多线程

Java 是支持多线程的语言。多线程可以带来更好的交换响应和实时行为。多线程是一种应用程序设计方法。线程是从大进程里分出来的、小的、独立的进程，使得在一个程序里可同时执行多项小任务。多线程带来的好处是具有更好的交互性能和实时控制性能，但采用传统的程序设计语言（如 C/C++）实现多线程非常困难。Java 实现了多线程技术，提供了一些简便地实现多线程的方法，并拥有一套高复杂性的同步机制。

11. 动态

Java 语言具有动态特性，能够适应不断发展的环境。Java 的动态特性是其面向对象设计方法的扩展，库中可以自由地添加新方法和实例变量，允许程序动态地装入运行过程中所需的类，Java 将符号引用信息在字节码中保存后传递给解释器，再由解释器在完成动态连接类后，将符号引用信息转换为数据偏移量。存储器生成的对象不在编译过程中确定，而是延迟到运行时由解释器确定。

1.1.3 Java 平台

Java 推出了 3 个领域的应用平台：标准版 Java 2 Platform Standard Edition（Java SE）、企业版 Java 2 Platform Enterprise Edition（Java EE）和微型版 Java 2 Platform Micro Edition（Java ME）。

Java SE 是各应用平台的基础。Java SE 可以分为 4 个主要部分：JVM、JRE、JDK 和 Java 语言。为了能运行 Java 程序，平台上必须安装有 Java 虚拟机（Java Virtual Machine，JVM）。JVM 包含在 Java 运行环境（Java SE Runtime Environment，JRE）和 Java 开发包（Java Development Kit，JDK）等 java 软件中。JDK 包括了 JRE 及开发过程中所需要的一些工具程序，如 Javac、Java、Appletviewer 等。安装 JRE 软件则可以运行 Java 程序；安装 JDK 软件则不但能运行 Java 程序，还可以编译开发 Java 程序。Java 语言只是 Java SE 的一部分。此外，Java 提供了庞大且功能强大的 API(Application Programming Interface)类库，可以使用这些 API 作为基础进行程序开发，而无需重复开发功能相同的组件。

Java EE 以 Java SE 为基础，定义了一系列的服务、API、协议等，适用于开发分布式、多层式（Multi-tiered）、以组件为基础、以 Web 为基础的应用程序。整个 Java EE 的体系是相当庞大的，比较常用的技术有 JSP、Servlet、Enterprise JavaBeans（EJB）、Java Remote Method Invocation（RMI）等。

Java ME 是 Java 平台版本中最小的一个，是作为小型数字设备上开发及部署应用程序的平台，如消费型电子产品、嵌入式系统等。

1.2 Java 语言开发环境

为了能运行 Java 程序，平台上必须安装有 Java 虚拟机（Java Virtual Machine，JVM）。JVM 包含在 Java 运行环境（Java SE Runtime Environment，JRE）和 Java 开发包（Java Development Kit，JDK）等 Java 软件中。JDK 包括了 JRE 及开发过程中所需要的一些工具程序，安装 JDK 软件则可以运行及编译开发 Java 程序。下面将介绍如何下载安装 JDK，并且介绍安装后的配置和编译方法。本小节还会介绍集成开发工具 MyEclipse 及 Tomcat 服务器的安装配置。

1.2.1 Jdk 下载及安装

JDK 是 Java 开发工具包。JDK 中包括 Java 编译器（javac）、打包工具（jar）、文档生成器（javadoc）、查错工具（jdb），以及完整的 JRE（Java Runtime Environment，Java 运行环境）。

JDK 一般有 3 种版本，如表 1.1 所示。

表 1.1 JDK 版本

名称	说明
SE（Java SE）	Standard Edition，标准版，主要用于开发 Java 桌面应用程序
EE（Java EE）	Enterprise Edition，企业版，使用这种 JDK 开发 Java EE 应用程序，用于 Web 方面
ME（Java ME）	Micro Edition，微型版，主要用于移动设备、嵌入式设备上的 Java 应用程序

在开发 Java 程序之前，先要在本机上安装 Java 程序开发工具包 JDK。安装好后的 JDK 具有 bin、demo、jre、lib、src 等子目录。在 Sun 公司的网站上有免费的 JDK 可供下载。下载及安装步骤如下所述。

（1）打开浏览器，在地址栏输入 Sun 公司的网址：http://java.sun.com/javase/downloads /index.jsp，进入下载页面，在该页面中选择合适的 Java 版本，如图 1.1 所示。

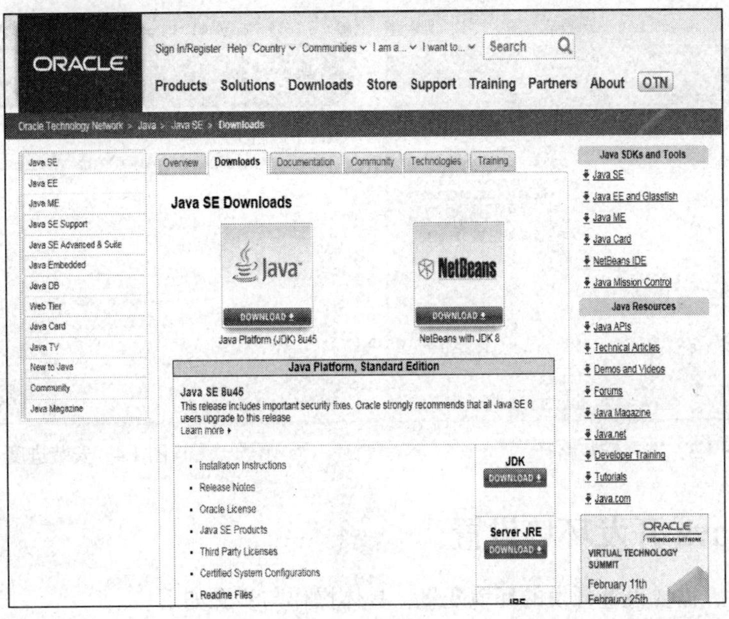

图 1.1　下载页面

（2）在图 1.2 所示的页面中，显示的是不同平台下的 JDK 安装包，如 Windows、Linux、Solaris。找到适合于自己计算机平台的 JDK 版本，这里下载 jdk-8u40-windows-i586.exe 可执行程序。需要注意的是，在下载 JDK 工具包之前，要选择接受协议。

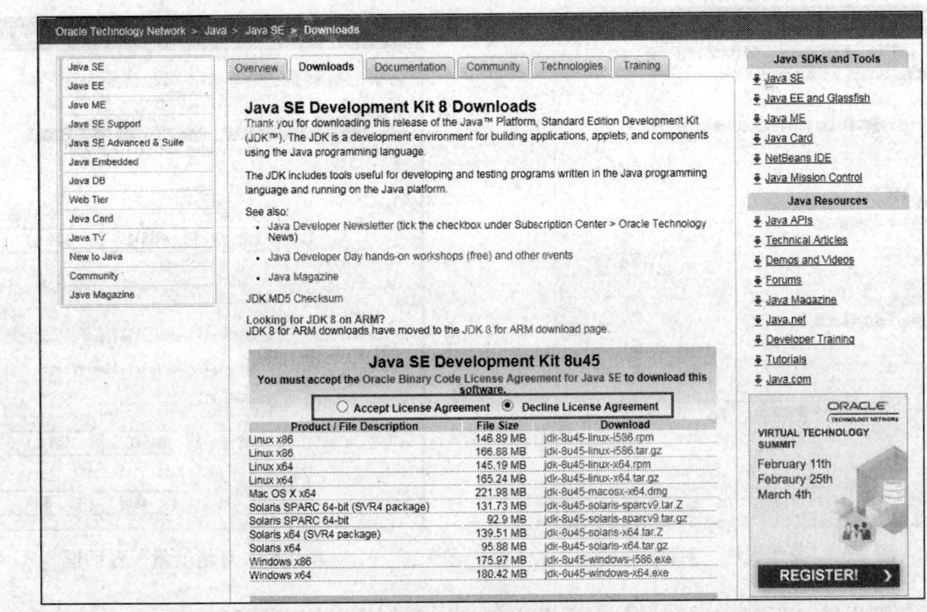

图 1.2　选择安装包

（3）下载好后双击图标，弹出安装向导对话框，在这里设置 JDK 的安装路径，可以自行更改路径，也可以使用默认路径，如图 1.3 所示。

（4）单击“下一步”按钮，开始安装 JDK，如图 1.4 所示。

（5）在安装过程中会出现安装 Java 运行环境的对话框，选择默认设置不做改变，单击“下一步”按钮，继续安装，安装成功后会出现相应的提示信息。

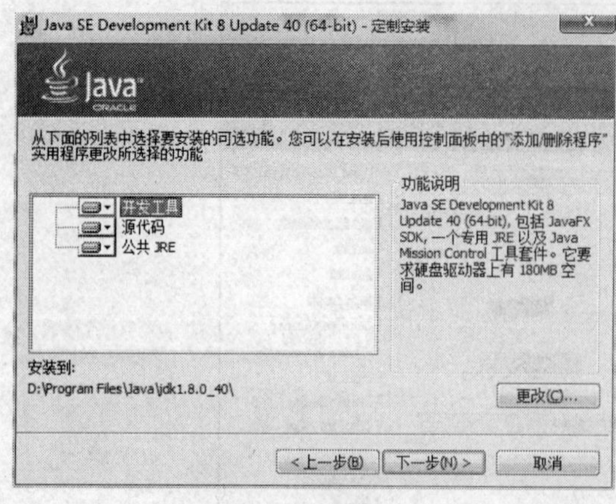

图 1.3　选择安装路径

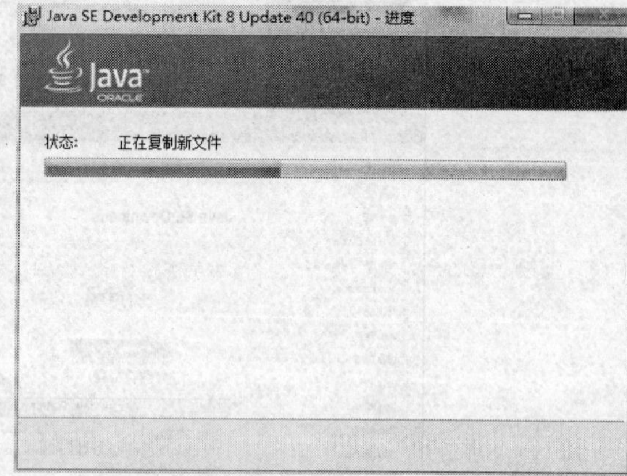

图 1.4　安装进度

1.2.2　Java 开发环境设置

JDK 安装后，需要设置开发的环境变量，具体操作与步骤如下。

（1）在“我的电脑”上单击鼠标右键，在弹出的快捷菜单中选择“属性”命令，弹出“系统属性”对话框，选择“高级”选项卡，如图 1.5 所示。

（2）单击“环境变量”按钮，在打开的“环境变量”对话框中，单击“新建”按钮，如图 1.6 所示。

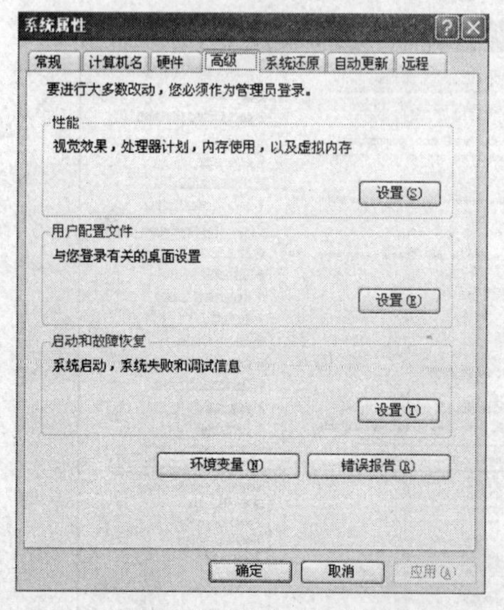

图 1.5　“系统属性”对话框

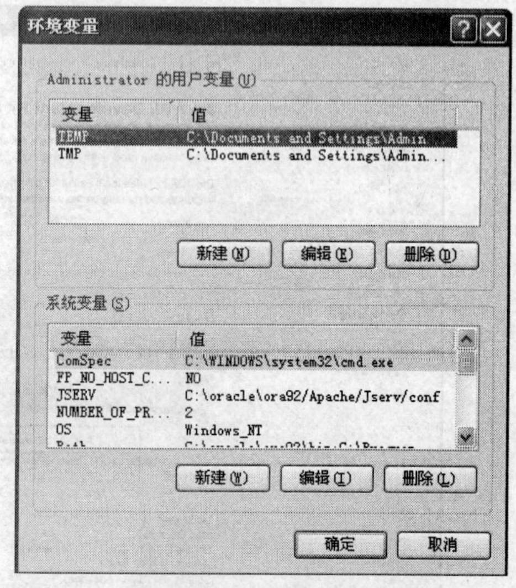

图 1.6　“环境变量”对话框

（3）在弹出的“新建用户变量”对话框中创建一个新的系统变量名“JAVA_HOME”，其值为 JDK 的安装路径。单击“确定”按钮保存，如图 1.7 所示。

图 1.7　新建用户变量

（4）在图 1.6 中选中“Path”变量名，单击“确定”按钮，在弹出的“编辑系统变量”对话框中设置变量值为“%JAVA_HOME%\bin”，单击“确定”按钮保存，如图 1.8 所示。

（5）新建“CLASSPATH”变量名，设置变量值为“%JAVA_HOME\lib\tools.jar”，单击“确定”按钮保存，如图 1.9 所示。

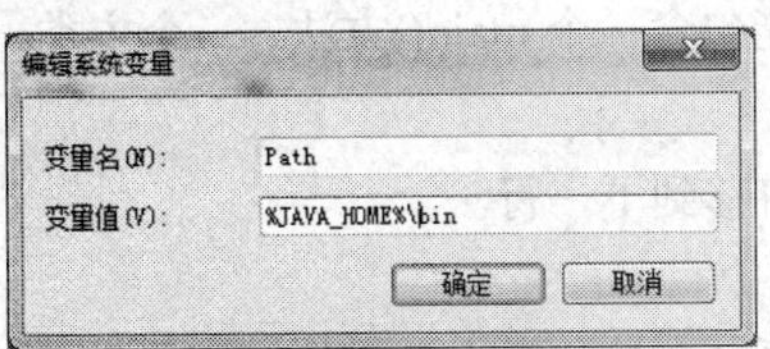

图 1.8　编辑系统变量

图 1.9　新建用户变量

（6）验证变量设置的正确性。选择“开始”→“运行”命令，在运行窗口中输入“cmd”命令，打开命令行编辑器窗口，在该窗口中分别输入“javac”（编译器）和“java”（解释器）命令，并按 Enter 键运行这两条命令。如果可以看到如下的帮助信息，则说明 JDK 的安装和设置是正确的，如图 1.10 所示。

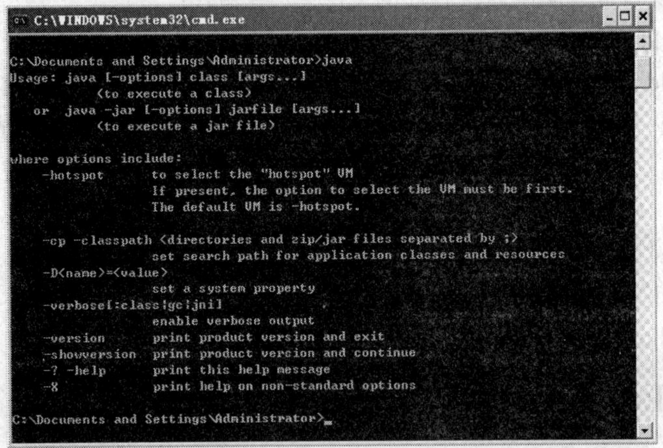

图 1.10　cmd 窗口

1.2.3　Java 程序的基本结构

用 Java 语言可以编写两种程序，一种是应用程序（Application），另一种是小应用程序（Applet）。应用程序是基于桌面型的应用，可以独立运行，也可以用在网络、多媒体等开发上。小应用程序是一种特殊的 Java 程序，不可以独立运行，是嵌入到 Web 网页中，且由带有 Java 插件的浏览器解释运行，主要用在 Internet 上。

下面所示是一段简单的 Java 应用程序 HelloWorld.java。

```
public class HelloWorld {    //类名要和文件名一致
    public static void main(String args[])
    {
      System.out.println("Hello World!");
    }    //结束main方法的定义
}    //结束类HelloWorld的定义
```

1. HelloWorld 程序中基本语法

（1）程序中关键词 class 定义类，其后大括号之间的语句构成了一个类体。

（2）关键词 public 表示"公有"，若类的对象或变量被定义为 public，表示该对象或变量可以被外界访问；若定义为 private，则不能被外界访问。

（3）main()方法又称为主方法，必须被说明为 public static void。包含 main()方法的类为主类，该程序中 HelloWorld 即为主类。一个 Java 应用程序只能包含一个 main()方法，一个主类，程序从主类的 main()方法开始执行。

（4）println()称为打印换行方法，显示文字后光标将移到下一行。

2. Java 程序的编译与运行

（1）编辑源程序。Java 源程序一般用 Java 作为扩展名，是一个文本文件，用 Java 语言写成，可以用任何文本编辑器创建与编辑。

例如，打开记事本编辑器，编辑 HelloWorld 代码，存放在特定文件夹。

（2）编译源程序。HelloWorld.java 程序不能直接运行，运行前使用"javac"编译器，读取 Java 源程序，并翻译成 Java 虚拟机能够明白的指令集合，并以字节码的形式保存在文件中。通常，字节码文件以 class 作为扩展名。

例如，在命令窗口进入保存 Java 程序的文件夹，并输入：

Javac HelloWorld.java

便将 HelloWorld.java 文件编译成 HelloWorld.class 类文件，并存放在当前文件夹。

（3）解释执行，使用"java"解释器，读取字节码，取出指令并且翻译成计算机能执行的代码，完成运行过程。字节码运行的平台是 Java 虚拟机，只要计算机上安装有 Java 虚拟机，不论采用哪种操作系统，硬件配置如何不同，运行的结果都一样。

例如，在命令窗口输入：

Java HelloWorld

便可以运行 HelloWorld.class 文件，得到程序结果。

1.2.4 MyEclipse 开发工具及使用

MyEclipse 是在 Eclipse 基础上加上插件开发而成的集成开发环境，主要用于 Java、Java EE 及移动应用的开发。单纯的 Eclipse 通常进行 Java Web 的开发时，需要安装 Eclipse 的插件，众多软件厂商和开源组织开发了相应的插件，其中 MyEclipse 最为常用。

1. 下载和安装 MyEclipse

下载 MyEclipse 可以通过官方网站链接 http://www.myeclipsecn.com/，也可以通过网络中的资源搜索，由于 MyEclipse 是一款非开源的开发工具，所以只能在官网中下载试用版本。图 1.11 给出了 MyEclipse Professional 2013 版本的下载安装，单击"Next"按钮进行安装。

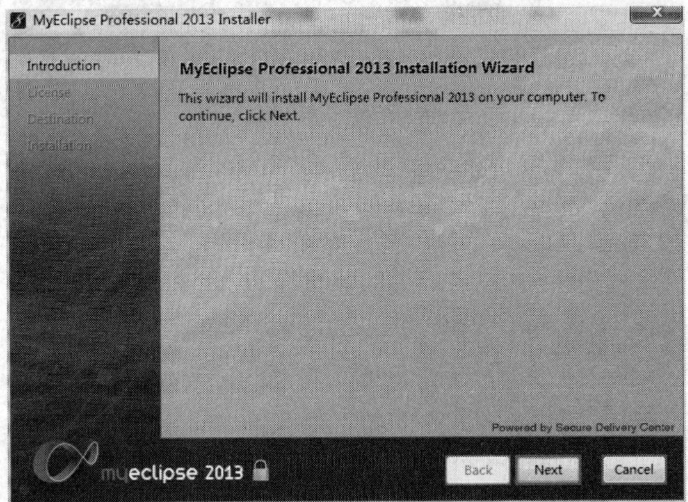

图 1.11　MyEclipse Professional 2013 安装

安装时需要手动选择工作空间目录，完成后启动 MyEclipse，工作界面如图 1.12 所示。

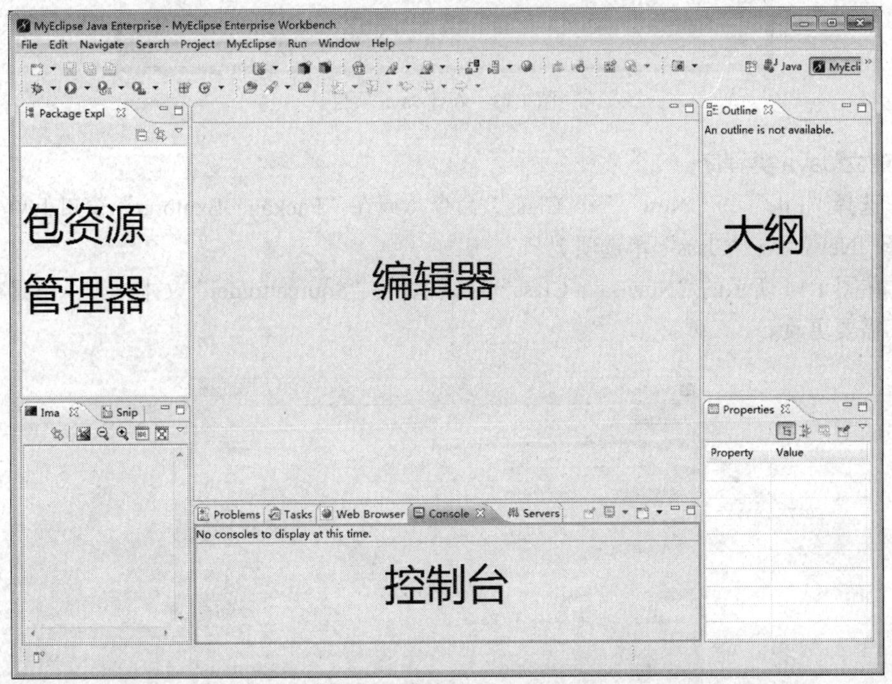

图 1.12　MyEclipse 工作界面

2. 建立 Java 项目

（1）选择“File”→“New”→“Project”命令（或是在“Package Explorer”窗口上单击鼠标右键，选择“New”→“Project”单选项）。

（2）在“New Project”对话框中，选择“Java Project”选项（或展开“Java”的数据夹，选择“Java Project”选项）。

（3）在“New Java Project”对话框中，输入 Project 的名称及创建 Project 位置（一般选择在工作空间中创建，然后单击“Finish”按钮，如图 1.13 所示。

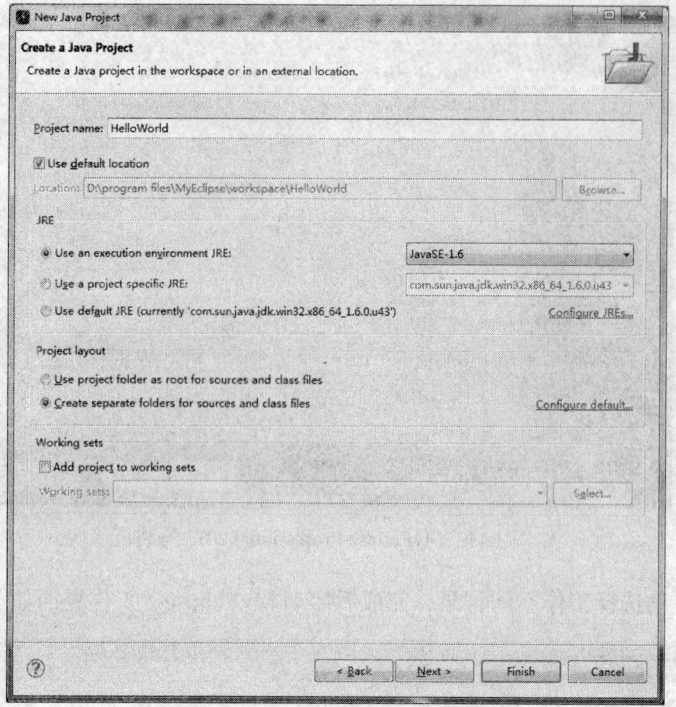

图 1.13　新建 Java 项目

3. 建立 Java 类程序

（1）选择“File”→“New”→“Class”命令（或在“Package Explorer”窗口上单击鼠标右键，选择“New”→“Class”单选项）。

（2）在图 1.14 所示的“New Java Class”对话框中，“Source folder”文本框默认值是项目的数据夹，不需要更改。

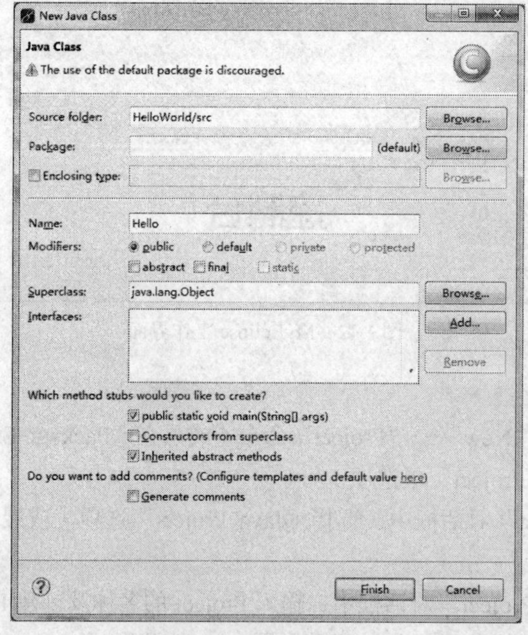

图 1.14　建立 Java 类

（3）在“Package”文本框输入程序套件的名称。

（4）在“Name”文本框输入Class Name。

（5）在“Which method Stubs would you like to create? ”选项中，如果勾选“public static void main(String[] args)”，在生成程序时会自动生成“main”方法。

（6）单击“Finish”按钮，会以套件新增适当的目录结构及Java 原始文件。

（7）在“main”方法中输入：“System.out.println（“Hello world !”）;”

4. 建立 Java 类程序

将当前编写的 Java 程序保存，选择“Run”→“Run”后，直接运行当前保存的程序，并将程序结果显示到 MyEclipse 控制台中，如图 1.15 所示。

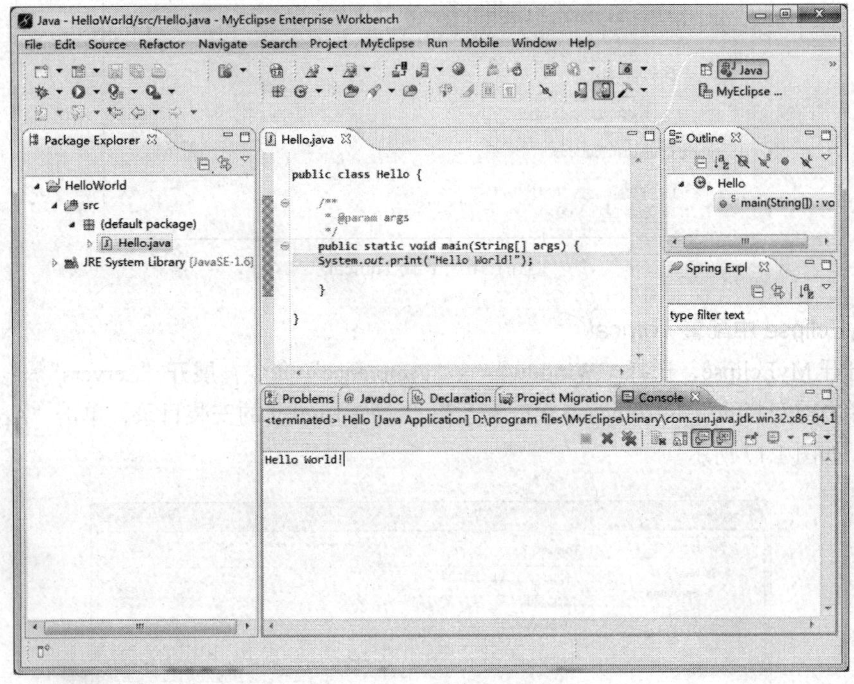

图 1.15　程序运行结果图

1.2.5　Tomcat 服务器的安装与配置

Tomcat 是 Apache 软件基金会的 Jakarta 项目中的一个核心项目，由 Apache、Sun 和其他一些公司及个人共同开发而成。Tomcat 服务器是一个免费的开放源代码的 Web 应用服务器，属于轻量级应用服务器，在中小型系统和并发访问用户不是很多的场合下被普遍使用。在 Java Web 的开发中，常将项目部署至 Tomcat 服务器上，访问项目的页面。

1. 下载 Tomcat 服务器

Tomcat 服务器可以通过 Apache 官方网站的 Tomcat 链接“http://tomcat.apache.org/”下载得到，例如下载 Tomcat 7.0 版本，如图 1.16 所示。

其中“Binary Distributions”下是编译好的二进制文件，“Source Code Distributions”下是 Tomcat 的源代码。这里选择“Core”下的 Windows 安装文件，注意区分 32 位和 64 位的区别。下载 zip 格式的 Tomcat 后，解压缩后安装完成，并记住 Tomcat 的安装目录，安装 Tomcat 前确保 JDK 已正确安装。

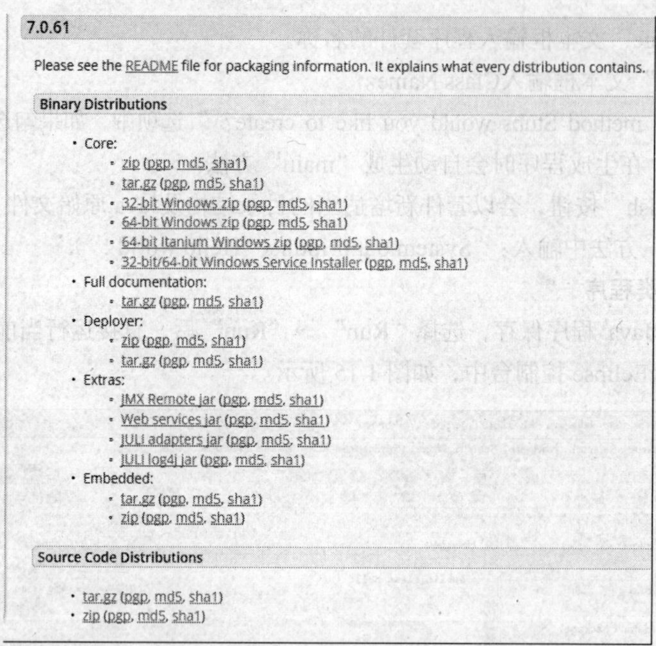

图 1.16　下载 Tomcat

2. MyEclipse 中配置 Tomcat

首先打开 MyEclipse，选择“Window”→“Preference”命令，展开“Servers”→“Tomcat”选项，选择之前安装的版本 Tomcat 7.x 进入系统，添加 Tomcat 的安装目录，单击“Apply”按钮完成操作，如图 1.17 所示。

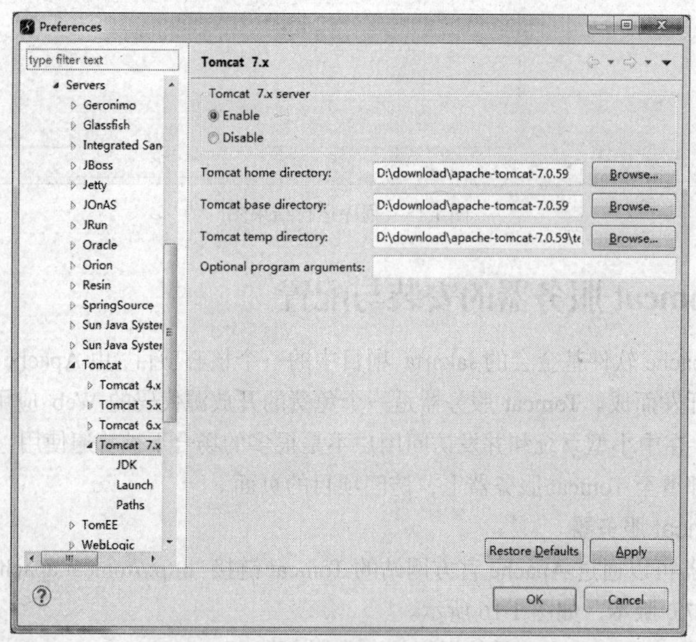

图 1.17　配置 Tomcat

此时 MyEclipse 中已成功配置 Tomcat，但在 MyEclipse 新建 Web 项目时，需要手动将项目添加至 Tomcat 目录。在左侧包资源管理器中右击项目名称，选择“MyEclipse”→“Add and Remove

Project Deployments”命令，选择项目的名称，单击“Add”按钮，添加配置的 Tomcat 版本，单击“Finish”按钮，提示成功，步骤如图 1.18、图 1.19 所示。

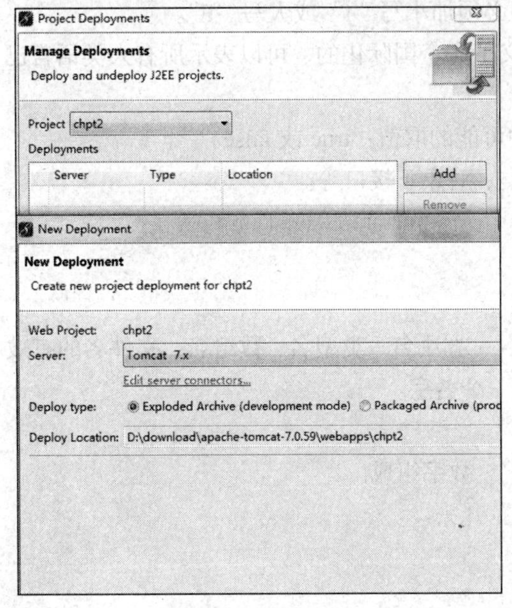

图 1.18　部署工程

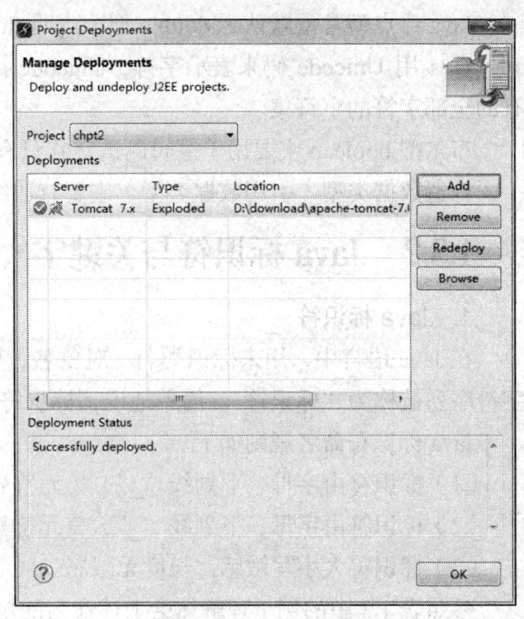

图 1.19　项目部署成功

1.3　Java 语言编程基础

Java 语言同样遵循着严格的编程规范，Java 的学习者必须了解 Java 语言的编程基础，遵循其编程规范，才能编写出好的 Java 程序。本节介绍的 Java 语法基础主要有 Java 的基本数据类型、标识符与关键字、运算符、控制语句与异常处理等。

1.3.1　Java 基本数据类型

Java 是一种强类型语言，它有着非常丰富的数据类型，可分为原始数据类型（基本数据类型）和构造数据类型（引用数据类型）两大类，其基本数据类型如表 1.2 所示。

表 1.2　Java 基本数据类型

数 据 类 型	关　键　字	分配内存空间（位宽）	取 值 范 围
字节型	byte	8	−128~127
短整型	short	16	-2^{15}~2^{15}-1
整型	int	32	-2^{31}~2^{31}-1
长整型	long	64	-2^{63}~2^{63}-1
单精度型	float	32	1.4E−45~3.4E+38
双精度型	double	64	4.9E−324~1.797E+308
字符型	char	16	Unicode 字符
布尔型	boolean	1	true、false

说明：

Java 语言的整型常量默认为 int 型，声明 long 型常量必须加小写“l”或大写“L”。

Java 浮点型常量默认为 double 型，声明 float 型必须加小写“f”或大写“F”。

Java 用 Unicode 码来表示字符，Unicode 码定义了完全国际化的、可以表示所有人类语言已有的全部字符的字符集。

布尔型 boolean 主要用于逻辑测试。它只有两种可能的取值：true 或 false。

构造数据类型（引用数据类型）主要有数组、类、对象、接口等。

1.3.2 Java 标识符与关键字

1. Java 标识符

在 Java 语言中，用来标识类名、对象名、变量名、方法名、类型名、数组名、文件名的有效字符序列被称为“标识符”。简单地说，标识符就是一个名字。

Java 标识符命名规则如下。

（1）标识符由字母、下划线“_”、美元符号“$”、数字组成。

（2）标识符由字母、下划线“_”、美元符号“$”开头。

（3）标识符大小写敏感，长度无限制。

参照表 1.3 中的例子理解体会上述命名的规则。

表 1.3 Java 标识符

合法的标识符	不合法的标识符
HelloWorld	class
TableTest	TableTest #
_123456	25.5
$ddf344	Hello world

2. Java 关键字

关键字就是 Java 中赋以特定的含义，并用作专门用途的单词，不能作为一般的标识符。程序员在编写程序时，不能再使用这些词汇来命名标识符，也不能改变这些词汇的含义。这些专有词汇，称为“关键字”，如表 1.4 所示。注意所有 Java 关键字都是小写英文。

表 1.4 Java 关键字

abstract	boolean	break	byte	case	catch	char	class
const	continue	default	do	double	else	extends	false
final	finally	float	for	goto	if	implements	import
instanceof	int	interface	long	native	new	null	package
private	protected	public	return	short	static	super	switch
synchronized	this	throw	throws	transient	true	try	void
volatile	while						

说明：

数据和返回类型：int、void、return。

包/类/接口：package、class、interface。

修饰符：public、protected、private、final、abstract、static。

控制：continue、break、if、else、switch、case、default、do、for。

异常：try、catch、throw、throws、finally。

运算符：new、instanceof。

1.3.3　运算符

对于数据进行的操作称为运算，表示各种不同运算的符号称为运算符，参与运算的数据称为操作数。表达式是由操作数和运算符按一定的语法形式组成的有意义的符号序列。对表达式中的操作数进行运算得到的结果称为表达式的值，表达式值的数据类型即为表达式的类型。表达式的值还可以用作其他运算的操作数，形成更复杂的表达式。

Java 语言中包括以下运算符。

单目算术运算符：+ +、- -、-。

双目算术运算符：+、-、*、/、%。

赋值运算符：=、+=、-=、*=、/=、%=、&=、|=、>>=、>>>=、<<=、^=。

关系运算符：>、<、>=、<=、==、!=。

逻辑运算符：&、&&、|、||、!、^。

位运算符：>>、<<、&、|、^、~。

条件运算符：?　:。

成员运算符：.。

下标运算符：[]。

实例类型判断运算符：instanceof。

创建对象运算符：new。

强制类型转换运算符：(类型名称)。

1. 单目算术运算符

单目运算符的操作数只有一个，算术运算符中有 3 个单目运算符，如表 1.5 所示。

表 1.5　单目算术运算符

运算符	运算	例子
++	自增	a++或++a
--	自减	a- -或- -a
-	求相反数	-a

2. 双目算术运算符

双目算术运算符如表 1.6 所示。

表 1.6　双目算术运算符

运算符	运算	例子	功能
+	加	a+b	a 与 b 相加的和
-	减	a-b	a 与 b 相减的差
*	乘	a*b	a 与 b 相乘的积
/	除	a/b	a 除以 b 的商
%	取余	a%b	a 除以 b 所得的余数

3. 关系运算符

关系运算符是比较两个数据的大小关系的运算符，关系运算符的结果是布尔型，即 true 或 false，如表 1.7 所示。

表 1.7　关系运算符

运 算 符	含　义	表 达 式	结　果
>	大于	43>13	true
<	小于	'a'<'f'	true
>=	大于等于	4.3>=1.3	true
<=	小于等于	'A'<=13	false
==	等于	'B'==66	true
!=	不等于	'A'!=65	false

4. 逻辑运算符

逻辑运算符是针对布尔型数据进行的运算，运算结果仍为布尔型量。常用的逻辑运算符如表 1.8 所示。

表 1.8　逻辑运算符

运 算 符	运　算	例　子	运 算 结 果
!	逻辑反	! x	x 真时为假，x 假时为真
||	逻辑或	x||y	x 和 y 都假时结果才为假
&&	逻辑与	x&&y	x 和 y 都真时结果才为真
^	布尔逻辑异或	x^y	x 和 y 不同值时结果为真
&	布尔逻辑与	x&y	x 和 y 都真时结果才为真
|	布尔逻辑或	x|y	x 和 y 都假时结果才为假

这里要特别对逻辑运算进行进一步说明：Java 提供了两个在其他大多数计算机语言中没有的有趣的布尔运算符。这就是逻辑与和逻辑或这两个运算符的特殊短路版本。例如，A && B、A || B。在逻辑或的运算中，如果第 1 个运算数 A 为真，则不管第 2 个运算数 B 是真是假，其运算结果都为真。同样，在逻辑与的运算中，如果第 1 个运算数 A 为假，则不管第 2 个运算数是真是假，其运算结果都为假。如果运用||和&&的形式，而不是|和&，那么一个运算数就能决定表达式的值，Java 的短路版本就不会对第 2 个运算数求值，只有在需要时才对第 2 个运算数求值。为了完成正确的功能，当右边的运算数取决于左边的运算数是真还是假时，短路版本是很有用的。例如，下面的程序语句说明了短路逻辑运算符的优点，用它来防止被 0 除的错误：

```
if (count != 0 && num / count > 10)
```

既然用了短路与运算符，就不会有当 count 为 0 时产生的意外运行错误。如果该行代码使用标准与运算符（&），则它将对两个运算数都求值，当出现被 0 除的情况时，就会产生运行错误。

5. 位运算符

位运算指的是对操作数以二进制为单位进行的运算，运算结果为整数。也就是说，将操作数转换为二进制表示形式，然后按位进行布尔运算，运算的结果也为二进制。位运算符及其运算规则如表 1.9 所示。

表 1.9　位运算符

<table>
<tr><th>运 算 符</th><th>运　　算</th><th>例　子</th><th>运 算 规 则</th></tr>
<tr><td>&</td><td>按位与运算符</td><td>x &a</td><td>求 x 和 a 各二进制位与</td></tr>
<tr><td>|</td><td>按位或运算符</td><td>x |a</td><td>求 x 和 a 各二进制位或</td></tr>
<tr><td>^</td><td>按位异或运算符</td><td>x^a</td><td>求 x 和 a 各二进制位异或</td></tr>
<tr><td>~</td><td>按位取反运算符</td><td>~ x</td><td>求 x 各二进制位取反</td></tr>
<tr><td><<</td><td>左移位运算符</td><td>x<<a</td><td>x 各二进制位左移 a 位</td></tr>
<tr><td>>></td><td>右移位运算符</td><td>x>>a</td><td>x 各二进制位右移 a 位</td></tr>
<tr><td>>>></td><td>不带符号的右移位运算符</td><td>x>>>a</td><td>x 各二进制位右移 a 位，左边的空位一律补填零</td></tr>
</table>

6. 赋值运算符

简单赋值运算符形式为“=”，Java 中还提供了复合赋值运算符，其形式为“运算符<op>=”。复合赋值运算符的含义如表 1.10 所示。

表 1.10　复合赋值运算符

<table>
<tr><th>运 算 符</th><th>例　　子</th><th>运 算 符</th><th>例　　子</th></tr>
<tr><td>+=</td><td>x+=a 等价于 x=x+a</td><td>|=</td><td>x|=a 等价于 x=x|a</td></tr>
<tr><td>-=</td><td>x−=a 等价于 x=x−a</td><td>^=</td><td>x^=a 等价于 x=x^a</td></tr>
<tr><td>*=</td><td>x*=a 等价于 x=x*a</td><td><<=</td><td>x<<=a 等价于 x=x<<a</td></tr>
<tr><td>/=</td><td>x/=a 等价于 x=x/a</td><td>>>=</td><td>x>>=a 等价于 x=x>>a</td></tr>
<tr><td>%=</td><td>x%=a 等价于 x=x%a</td><td>>>>=</td><td>x>>>=a 等价于 x=x>>>a</td></tr>
<tr><td>&=</td><td>x&=a 等价于 x=x&a</td><td></td><td></td></tr>
</table>

7. 条件运算符

Java 中唯一的一个三元运算符是“?:”，它有 3 个操作数。这种语法的作用与双分支的选择语句很相似，但其返回值更直接，书写形式更简洁。其语法形式如下：

```
逻辑表达式? 表达式 1：表达式 2
```

其中，逻辑表达式为 boolean 类型表达式，先计算逻辑表达式的值，若为 true，则整个三目运算的结果为表达式 1 的值，否则整个运算结果为表达式 2 的值。

8. 运算符优先级

当一个表达式中有多个运算符参与混合运算时，表达式的运算次序取决于表达式中各种运算符的优先级，也就是说，不同的运算符有不同的优先级。Java 运算符优先级如表 1.11 所示。

表 1.11　运算符优先级

描　述	优 先 级	结 合 方 向
最高优先级	.、[ ]、( )	从左向右
单目运算	++、−−、！、~	从左向右
算数乘除运算	*、/、%	从左向右
算数加减运算	+、−	从左向右
移位运算	<<、>>、>>>	从左向右
关系运算	<、>、>=、<=	从左向右
相等关系运算	==、!=	从左向右

续表

描　述	优 先 级	结 合 方 向
按位与、布尔逻辑与	&	从左向右
按位异或	^	从左向右
逻辑与	&&	从左向右
逻辑或	||	从左向右
三元运算符	?　:	从右向左
赋值运算符	=和复合赋值运算符	从右向左

1.3.4　程序控制语句

Java 语言中的程序流程控制语句有 3 种：顺序结构、选择结构和循环结构。Java 的每条语句一般以分号（“;”）作为结束标志。

顺序结构：3 种结构中最简单的一种，即语句按照书写的顺序依次执行。

选择结构：又称分支结构，将根据布尔值来判断应选择执行哪一个流程分支。

循环结构：在一定条件下反复执行一段语句的流程结构。

1. If-else 选择语句

（1）简单的 if 选择语句

① if（条件表达式）。

```
语句                //只有一条语句
```

② if（条件表达式）。

```
{
    一条或多条语句      //多条语句必须加大括号
}
```

（2）if-else 双分支选择语句

① if（条件表达式）。

```
    语句 1
else
    语句 2
```

② if（条件表达式）。

```
{
    语句块 1
}
else
{
    语句块 2
}
```

（3）if-else-if 多分支选择语句

```
if（条件表达式 1）
{   语句块 1  }
else  if（条件表达式 2）
```

```
{   语句块 2   }
…
else  if（条件表达式 n-1）
{   语句块 n-1   }
else
{     语句块 n     }
```

2. switch 语句

当选择结构的分支越多时，if-else-if 语句就会变得越来越难以看懂。Java 提供了另一种多分支语句——switch 语句。switch 语句是多分支的开关语句。它的语法格式如下：

```
switch（表达式）{
      case 常量表达式 1：语句组 1;
                                   [break;]
      case 常量表达式 2：语句组 2;
                                   [break;]
      case 常量表达式 3：语句组 3;
                                   [break;]
      …
      case 常量表达式 n-1：语句组 n-1;
                                   [break;]
      default：语句块 n;
}
```

说明：

（1）switch 后面的表达式的值的类型可以是 char、byte、short、int 型，但不能是 boolean、long、float、double 型。

（2）case 后面的常量表达式的值的类型必须与 switch 后面的表达式的值的类型相匹配，而且必须是常量表达式或直接字面量。

（3）break 语句可以省略，但省略时应注意，此时程序将按照顺序逐个执行 switch 中的每一条语句，直到遇到右大括号或者 break 语句为止。

（4）语句块可以是一条语句，也可以是多条语句，但此处的多条语句不需要使用大括号。

（5）case 分支语句和 default 语句都不是必须的，也就是说它们在程序需要时是可以省略的。

3. while 循环语句

while 循环先判断条件后做循环。while 循环语句的语法格式如下：

```
while（布尔表达式）
{   语句块 //循环体    }
```

说明：

（1）表达式必须是任何运算结果为布尔值的表达式。

（2）只有一条语句时可以省略大括号，但不建议这么做。

（3）while 后面一定没有分号，在大括号后面也没有分号，这点初学者很容易忽略。

例 1-1　使用 while 语句编写程序，求 1000 以内的所有偶数的和（包含 1000）。

```
public class TestWhile{
  public static void main(String[] args){
     int i = 1, sum=0;
     while(i <= 1000){
```

```
        if(i%2==0)
            sum = sum + i;
        i++;
    }
    System.out.println("1000 之内所有偶数的和是：" + sum);
  }
}
```

4. do-while 循环语句

与 while 语句功能相似的另一个循环语句是 do-while 语句。先做循环后判断条件。

do-while 语句的语法格式如下：

```
do
{ 语句或语句块 //循环体 }
while（布尔表达式）;
```

说明：

（1）表达式必须是任何计算结果为布尔值的表达式。

（2）只有一条语句时可以省略大括号，但不建议这么做。

（3）while 后面一定要有分号，以表示 do-while 语句的结束。这一点与 while 语句不同。

例 1-2 使用 do-while 语句编写程序，求 1000 以内的所有偶数的和（包含 1000）。

```
public class TestDoWhile{
  public static void main(String[] args){
    int i = 1, sum=0;
    do{
        if(i%2==0)          //判断 i 是否为偶数
            sum = sum + i;
        i++;
    } while(i <= 1000);
    System.out.println("1000 之内所有偶数的和是：" + sum);
  }
}
```

5. for 循环语句

在循环次数已知的情况下，可以使用 for 语句替代 while 或 do-while 语句。其语法格式如下：

```
for（初始化表达式；条件表达式；迭代式）
{   循环体语句  }
```

说明：

（1）初始化表达式通常用来对循环变量进行初始化，或者定义循环变量并初始化，在循环过程中只会被执行一次。

（2）条件表达式是一个布尔表达式，即运算结果为布尔值的表达式。只有当条件为 true 时才会执行。

（3）迭代表达式用于改变循环条件的语句，为继续执行循环体语句做准备。

（4）初始化循环变量可以在 for 语句之前声明，此时 for 语句中的初始化表达式就可以省略；循环变量的迭代式也可以在 for 循环体内执行，如 for（;;）{…}。

例 1-3 使用 for 循环语句，求 1000 以内的所有偶数的和（包含 1000）。

```
public class TestFor{
  public static void main(String[] args){
```

```
    int  sum=0;
    for(int i = 1;i <=1000;i++){
        if(i%2==0)                    //判断 i 是否为偶数
            sum = sum + i;
    }
    System.out.println("1000之内所有偶数的和是：" + sum);
  }
}
```

6. 跳转语句

（1）break 语句在前面讲过的 switch 语句、循环语句中，可以使用 break 语句来终止 switch 语句、循环语句的执行，而整个程序继续执行后面的语句。

（2）continue 语句：continue 语句的作用是提前结束本次循环，立即开始下一轮循环。

continue 语句与 break 语句的区别：continue 语句只结束本次循环，并不终止整个循环的执行；而 break 语句则是结束整个循环过程，不再判断循环的条件是否成立。

1.3.5 Java 异常处理

异常是指程序在运行过程中出现由于硬件设备问题、软件设计错误等导致的程序异常事件，Java 异常是一个描述在代码段中发生的异常情况的对象。

Java 异常处理通过 5 个关键字控制：try、catch、throw、throws 和 finally。异常处理可以有以下几种。

（1）对运行时异常不做处理，由系统寻找处理的代码。

（2）使用 try-catch-finally 语句捕获异常。

（3）通过 throws 子句声明异常，还可自定义异常，用 throw 语句抛出。

1. try 语句

为防止和处理一个运行时错误，只需要把你所要监控的代码放进一个 try 块就可以了。

例 1-4　利用 try 语句捕获异常。

```
try{
   int n;
   n = new classTest.method();
   int[] array = new int[10];
   for (int i=0; i<n ; i++){
      array[i] =i ;                                   //可能引起数组越界异常
    }

}
```

try 语句可以被嵌套。一个 try 语句可以在另一个 try 块内部。每次进入 try 语句，异常的前后关系都会入堆栈。如果一个内部的 try 语句不含特殊异常的 catch 处理程序，堆栈将弹出，下一个 try 语句的 catch 处理程序将检查是否与之匹配。这个过程将继续直到一个 catch 语句匹配成功，或者是直到所有的嵌套 try 语句被检查耗尽。如果没有 catch 语句匹配，Java 运行时系统将处理这个异常。

2. catch 语句

紧跟着 try 块的，包括一个说明你希望捕获的错误类型的 catch 子句。catch 语句和 try 语句一起构成异常测试和捕获处理代码形式。

```
try-catch 异常处理代码块的基本形式：
try{
```

```
        //监视可能发生异常的代码块
    }
    catch（异常类型异常对象名）  //捕获并处理异常
    {
        //异常处理代码块
    }
```

改写上面的程序，包含一个因为被零除而产生的 ArithmeticException 异常的 try 块和一个 catch 子句。

例 1-5 利用 try-catch 语句捕获 ArithmeticException 异常。

```
Public class Test
{
    public static void main(String args[])
    {
        int a, b;
        try
        {
            a = 0;
            b= 5 / a;
            System.out.println("This will not be printed.");
        }
        catch (ArithmeticException e)
        {
            System.out.println("Division by zero.");
        }
     System.out.println("After catch statement.");
    }
}
```

该程序输出如下：

```
Division by zero.
After catch statement.
```

某些情况，由单个代码段可能引起多个异常。处理这种情况，可以定义两个或更多的 catch 子句，每个子句捕获一种类型的异常。当异常被引发时，每一个 catch 子句被依次检查，第一个匹配异常类型的子句被执行。当一个 catch 语句执行以后，其他的子句被旁路，执行从 try-catch 块以后的代码开始继续。多个 catch 异常处理代码块的基本形式：

```
try{
    //可能发生异常的代码块
}catch（异常类型 1 异常对象名 1）{
    //异常处理代码块 1
}
…
catch（异常类型 n 异常对象名 n）{
    //异常处理代码块 n;
}
```

3. finally 语句

finally 关键字是 Java 异常处理的最后一个语句，无论 try 语句中是否出现异常、出现哪种类型异常，finally 关键字中包含的语句都必须被执行。如果异常被引发，finally 甚至是在没有与该异常相匹配的 catch 子句的情况下也将执行。finally 子句是可选项，可以有也可以无。然而每一个

try 语句至少需要一个 catch 或 finally 子句。

下面的例子显示了 3 种不同的退出方法，每一个都执行了 finally 子句。

例 1-6　利用 try-finally 捕获异常。

```
class FinallyDemoTest
{
    static void procA()
    {
    try
    {
    System.out.println("inside procA");
    throw new RuntimeException("demo");
    }
    finally
    {
    System.out.println("procA's finally");
        }
    }
    static void procB()
    {
    try
    {
            System.out.println("inside procB");
            return;
    }
    finally
    {
            System.out.println("procB's finally");
    }
}
static void procC()
{
    try
    {
        System.out.println("inside procC");
    }
    finally
    {
        System.out.println("procC's finally");
    }
   }
public static void main(String args[])
{
    try
    {
        procA();
    }
    catch (Exception e)
    {
        System.out.println("Exception caught");
        }
        procB();
        procC();
    }
}
```

procA()中 finally 子句在退出时执行。procB()中返回之前 finally 子句执行。procC()中没有错误，但 finally 块仍将执行。

下面是上述程序的输出结果：

```
inside procA
procA's finally
Exception caught
inside procB
procB's finally
inside procC
procC's finally
```

思考与习题

1. 填空题：Java 语言中的程序流程控制语句有 3 种：________、________和________。
2. 填空题：Java 异常处理通过 5 个关键字控制：try、________、throw、throws 和________。
3. 简答题：简述 Java 的特点及平台。
4. 编程题：安装并配置 JDK、MyEclipse、Tomcat。
5. 编程题：求两个数的最大公约数和最小公倍数。
6. 编程题：对 10 个整数的排序，用冒泡排序算法实现。

第2章 Java 面向对象技术基础

Java 语言作为一种面向对象语言，其将解决的问题分解成各个对象，建立对象的目的不是为了完成一个步骤，而是为了描述解决问题的各个步骤中的行为。在面向对象的编程中，程序将围绕着被操作的对象来设计，而不是操作本身。面向对象编程的三大特点是封装、继承与多态性。

本章将具体介绍 Java 的面向对象技术，主要包括类、对象、方法，以及面向对象的特点等。

2.1 类及对象

2.1.1 面向对象基本概念

传统的高级程序设计语言，如 Pascal 语言、C 语言等，都是面向过程的语言。该程序设计思想中，是以功能或模块作为问题域的解决方法，将存放基本数据类型的变量作为程序处理对象，抽象简单，信息暴露，在程序设计中，采用过程或函数来实现有限的代码重用，多个过程或函数相互调用，共同完成统一的功能，当项目较大时，程序调试和维护将变得异常困难。

为了解决面向过程设计语言的缺陷，面向对象的编程思想应运而生，即以信息隐藏和数据抽象概念为基础，以简单、直观、接近人类的自然思维方式来解决问题域内容，划分并构建问题域中的对象，以及描述对象的状态和行为。面向对象程序设计的优点有：开发时间短、效率高、可靠性高、开发程序健壮、应用程序易于维护、易于更新升级。

对象是系统中描述客观事物的一个实体，我们所见到的任何一个客观事物都可以看作是一个对象，如一张桌子、一面镜子、一台计算机等。对象是构成系统的一个基本单位，由一组属性和对这组属性进行操作的一组服务组成。属性是用来描述对象静态特征的一个数据项，服务是用来描述对象动态特征的一个操作序列，对象是属性和服务的结合体。

Java 是纯面向对象语言，核心是面向对象。面向对象具有封装、继承、多态 3 个特性。

封装是把对象的属性和服务结合成一个独立的系统单位，它是面向对象的一个重要概念。封装是一种信息隐蔽技术，用户只能见到对象封装界面上的信息，对象内部对用户是隐蔽的。也就是说，用户只知道某对象是“做什么”的，而不知道“怎么做”。封装将外部接口与内部实现分离开来，用户不必知道行为实现的细节，只需用消息来访问该对象。封装体现了良好的模块性，它将定义模块和实现模块分开。封装使对象的内部软件的范围边界清楚，使模块内部的数据受到很好的保护，避免外部的干扰。封装大大增强了软件的可维护性，这也是软件技术追求的目标。

继承是一个对象获得另一个对象的属性的过程，对象类之间的层次关系的内涵即为继承。继承支持了按层分类的概念。通过继承，子类可以自动拥有父类的全部属性和服务。当类 A 不但具有类 B 的属性，还具有自己的独特属性时，称类 A 继承了类 B，继承关系常称为“即是”关系。继承分为单继承和多重继承两种。在类的层次结构中，一个类可以有多个子类，也可以有多个超类。如果一个类至多只能有一个超类，则一个类至多只能直接继承一个类，这种继承方式称为单继承。如果一个类可以直接继承多个类，这种方式称为多重继承。如图 2.1 所示。

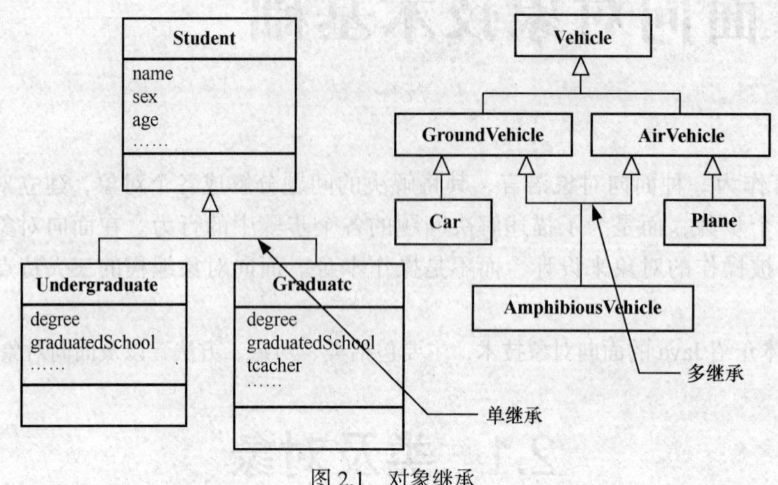

图 2.1　对象继承

多态性是允许一个接口被多个同类动作使用的特性，如在一个类中可以定义多个具有相同名的方法，或在子类中定义和父类中相同名字的方法，具体使用哪个动作与应用场合有关。多态性的概念经常被说成是“一个接口，多种方法”，这意味着可以为一组相关的动作设计一个通用的接口。选择应用于每一种情形的特定的动作（即方法）是编译器的任务，程序员无须手工进行选择，只需记住并且使用通用接口即可。多态性为程序设计提供了灵活性，减少了信息冗余，提高了软件可复用性及可扩充性。

2.1.2　类的定义

在面向对象设计中，首先要定义类。将数据抽象和对数据进行操作的方法结合在一起的完整定义称为类。类是 Java 的核心和本质。它是 Java 语言的基础。类是基于对象的基本概念，类就是对象的模板（template），而对象就是类的一个实例（instance）。一个类定义了将被一个对象集共享的结构和行为。一个给定类的每个对象实例都包含这个类定义的行为和结构。所以，类是一种逻辑结构，而对象是真正存在的物理实体（也就是对象占用内存空间）。

Java 是纯面向对象语言，即所有的代码都封装在类中。对一个类进行定义时，必须明确声明类的属性和方法。

使用关键字class 来声明类。

类的一般定义形式为：

```
[修饰符] class　类名称 [extends　父类名称] [implements　接口名称，……]
{
    ……//成员变量
    ……//成员方法
}
```

在该类定义中，有类的一些属性内容，如类的继承、类实现的接口、类的访问权限、类的抽象性与否等。

修饰符包括以下内容。

public：表示该类可以被所有的其他类使用。

protected：表示该类只能作内部类即嵌套类，同时可以被同一包或该类的派生类使用。

private：表示该类只能作内部类即嵌套类，并且只能被包含在它的外部类使用。

默认：表示该类可以被同一包的其他类使用。定义类时不写修饰符就是默认。

final：表示该类不能为父类。即无法从该类派生类。

abstract：表示该类为抽象类。即无法产生该类的实例。

static：表示为静态类，这个修饰符只能用于内部类即嵌套类，此时表示内部类的对象不属于外部类对象而独立存在。

Extends：表示直接继承的父类。

Implements：表示类实现的接口。

例 2-1 定义一个名为student 类的格式。

```
class student
{
        //定义属性或称实例变量
        public String sno ;  //学号
        public String sname ;  //姓名
        //……
        //定义方法
        public String getName(String name)
        {
            //方法体
        }
        …
}
```

2.1.3 对象的创建和使用

面向对象的编程思想是每个事物都是对象。类是对现实世界的抽象，类定义实现了对客观事物的描述，可以看作是创建一种新的数据类型，可以使用这种类型来声明该种类型的对象。要获得一个类的对象，必须声明该类类型的一个变量，同时，该声明要创建一个对象的实际的物理拷贝，并把对于该对象的引用赋予该变量。Java 语言提供了创建实例对象的操作符，这是通过使用 new 运算符实现的。new 运算符为对象动态分配内存空间，并返回对它的一个引用。当程序不再需要使用该对象时，通常会调用内存垃圾处理程序来进行清理。

例 2-2 定义 Student 类的程序。

```
class  Student //定义名为 Student 的类
{
    String sno;
    String sname;
}
class  StudentDemo    //定义名为 StudentDemo 的类
{
```

```
    public  static  void  main(String  args[])
  {
        Student s1 = new Student ();   //声明一个 Student 类型的对象 s1
        s1. sno = “B01010101”;
       s1.sname = “郑成功”;
       System.out.println("The student  is " + s1. sno + s1.sname );
      }
  }
```

Student s1= new Student ()语句创建了 Student 类的一个实例对象 s1，sno、sname 是在类体中声明的对象，称为成员对象。要访问类的成员变量，需使用点号“.”运算符。

可将 Student s1= new Student ()语句改写为下面的形式：

```
Student s1;
s1 = new Student ();
```

第 1 行声明了 s1，把它作为对于 Student 类型的对象的引用。当本句执行后，s1 包含的值为 null，表示它没有引用对象。这时任何引用 s1 的尝试都将导致一个编译错误。第 2 行创建了一个实际的对象，并把对于它的引用赋予 s1。现在，可以把 s1 作为 Student 的对象来使用，但实际上，s1 仅仅保存实际的 Student 对象的内存地址。

2.1.4 构造方法

类的方法实现了类所具有的行为，其他对象可以通过类的方法来对类进行访问。方法是具有一个名字的一段单独执行的代码段。

方法的一般形式：

```
[修饰符]  返回类型  方法名称 (参数列表)
{
    //  方法体
}
```

方法的修饰符为说明方法可见性的关键字，包括以下内容。

public：公共的、全局的，该方法可以被其他任何类使用。

protected：受保护的，不同包中的子类可以使用。

private：私有的，局部的，在同一类中使用。

默认：包访问，即同一包中的其他类可以调用。

final：终止方法，即该方法所在类的派生类是不能修改这个方法的。

abstract：抽象方法，相当于 C++中的纯虚方法。

static：静态方法，表示这个方法不需要对象来调用，是通过类来调用的，即该方法属于类。调用的格式为：

```
类名称.方法名称（实参列表）;
```

synchronized：表示该方法为同步方法。有关这方面的内容，将在多线程中进一步解释。

其中表示方法访问权限的修饰符有 4 种：public、protected、private、默认。

修饰符的组合使用规则如下：

public、private、protected 不能同时使用，它们是表示方法的访问权限；表示访问权限的修饰符可以与 final、static、final、abstract、synchronize 组合使用，如 public static final synchronized void method1()。

方法返回的数据类型应该是任何合法有效的类型，包括创建的类的类型。如果该方法无返回值，则它的返回必须为 void。方法名可以是任何合法的标识符。

参数列表是一系列类型和标识符对，用逗号分开。参数本质上是变量，它接收方法被调用时传递给方法的参数值。如果方法没有自变量，那么自变量列表就为空。对不是 void 返回值的方法，使用 return 语句：return value。

若每次创建实例变量都要初始化类中的所有变量，工作是重复而乏味的。如果一个对象被创建时就把对它的属性变量值设置好，则程序将更简单明了。在 Java 语言中，通过使用构造方法来完成实例变量值初始化。

构造方法与它的类同名，它的语法与方法类似。一旦定义了构造方法，在对象创建后，在 new 运算符完成前，构造方法立即自动调用。构造方法没有任何返回值。构造方法的任务就是初始化一个对象的内部状态，以便使创建的实例变量能够完全初始化，可以被对象使用。

例 2-3 定义一个 Car 类，定义构造方法来初始化该汽车的状态属性。

```
class Car
{
    String color;
    double weight;
    Car() //构造方法
    {
        color = "red";
        weight = 125.1;
         System.out.println("Constructing Car");
      }
      String printColor()
      {
         return color ;
      }
}
class CarDemo
{
      public static void main(String args[])
      {
            Car car1 = new Car ();
            Car car2 = new Car ();
            String c;
            c = car1.printColor();
            System.out.println("car1 color is " + c);
            c = car2.printColor();
            System.out.println("car2 color is " + c);
      }
}
```

该程序的输出如下：

```
Constructing Car
Constructing Car
car1 color is red
car2 color is red
```

例 2-4 定义一个 Car 类，定义一个自变量构造方法初始化类的属性。

```
class Car
{
```

```
    String color;
  double weight;
  Car() //构造方法
  {
    color = "red";
    weight = 125.1;
    System.out.println("Constructing Car");
  }
  Car(String c, double w)  //构造方法重载
  {
    color = c;
    weight = w;
    System.out.println("Constructing Car");
  }
  String printColor()
  {
       return color ;
  }
  }
  class CarDemo
  {
    public static void main(String args[])
    {
      Car car1 = new Car ("green", 10);
      Car car2 = new Car ("blue", 5);
      String c;
      c = car1.printColor();
      System.out.println("car1 color is " + c);
      c = car2.printColor();
      System.out.println("car2 color is " + c);
    }
  }
```

该程序的输出如下：

```
Constructing Car
Constructing Car
car1 color is green
car2 color is blue
```

2.1.5 方法重载

方法重载是在一个类中定义多个相同方法名的方法，只是该多个方法的参数数量、类型及返回类型不同。在这种情况下，该方法就被称为重载（overloaded）。方法重载是 Java 实现多态性的一种方式。当一个重载方法被调用时，Java 用参数的类型和（或）数量来表明实际调用的重载方法的版本。当 Java 调用一个重载方法时，参数与调用参数匹配的方法被执行。

例 2-5 一个说明方法重载的简单例子。

```
class OverloadTest
{
   void test()
   {
      System.out.println("No parameters");
```

```
    }
    void test(int a)
    {
        System.out.println("a: " + a);
    }
    void test(int a, int b)
    {
        System.out.println("a and b: " + a + " " + b);
    }
    double test(double a)
    {
        System.out.println("double a: " + a);
        return a*a;
    }
}
class Overload
{
    public static void main(String args[])
    {
        OverloadTest ot = new OverloadTest();
        double result;
        ot.test();
        ot.test(1);
        ot.test(1, 2);
        result = ot.test(0.5);
        System.out.println("Result of ot.test(0.5): " + result);
    }
}
```

该程序的输出如下：

```
No parameters
a: 1
a and b: 1 2
double a: 0.5
Result of ot.test(0.5): 0.25
```

从上述程序可见，test()被重载了 4 次。第 1 个版本没有参数，第 2 个版本有一个整型参数，第 3 个版本有两个整型参数，第 4 个版本有一个 double 型参数。

当一个重载的方法被调用时，Java 的自动类型转换也适用于重载方法的自变量。

例 2-6 重载方法的自变量可以转换类型。

```
class OverloadTest
{
    void test()
    {
        System.out.println("No parameters");
    }
    void test(int a, int b)
    {
        System.out.println("a and b: " + a + " " + b);
    }
    void test(double a)
    {
```

```
        System.out.println("Inside test(double) a: " + a);
    }
}
class Overload
{
    public static void main(String args[])
    {
        OverloadTest ob = new OverloadTest ();
        int i = 10;
        ob.test();
        ob.test(10, 20);
        ob.test(i);
        ob.test(1.2);
    }
}
```

该程序的输出如下：

```
No parameters
a and b: 10 20
Inside test(double) a: 10
Inside test(double) a: 1.2
```

在本例中，语句 ob.test(i)执行时找不到和它匹配的方法。但是，Java 可以自动地将整数转换为 double 型，然后调用 test(double)。

2.1.6 类的成员和关键字 this

1. 成员变量

类体中定义的变量为成员变量，而类的方法中定义的变量为局部变量；类的成员变量表明类的状态，完整的成员变量定义的形式为：

[修饰符] 数据类型变量名称；

修饰符表明该成员变量的对外可见性及特征，包括以下内容。

public：公共的、全局的，该变量可以被其他任何类使用。

protected：受保护的，不同包中的子类可以使用。

private：私有的，局部的，在同一类中使用。

默认：包访问，即同一包中的其他类可以调用。

final：终止属性，表示该属性值不能被改变，即是常量，通常用大写表示，如 private final double PI = 3.1415。

static：静态属性。表示这个属性不需要对象来调用，是通过类来调用的，即该属性属于类。调用的格式为：

```
类名称.属性名称
```

例如，private static int MaxSize=100；定义了一个静态的整型属性，并赋了初值。这个属性是该类中的所有对象共享的。

其中表示属性访问权限的修饰符有 4 种：public、protected、private、默认。

2. 关键字 this

this 关键字表明对象自身的引用，且可以在引用当前对象的所有方法内使用。如果类体中的某个成员变量与方法中的某个局部变量的名称相同，可以使用 this 关键字来存取同名的实例变量。

例 2-7　this 关键字存取同名的实例变量。

```
Box(double width, double height, double depth)
{
        this.width = width;
        this.height = height;
        this.depth = depth;
}
```

这个例子中，局部变量 width、height、depth 和类的成员变量命名相同，因此，使用this 关键字来存取同名的实例变量。

还可以用 this 来作为参数指明对象本身。

例 2-8　this 关键字指明对象本身。

```
class demo {
    void test (){
        otherclass oc = new otherclass();
        oc.othermethod(this); //使用 this 将对象 demo 传递给对象 oc 的方法 othermethod
}
}
```

2.2　封装、继承与多态性

面向对象是一种新兴的程序设计方法，或称为程序设计规范，其基本思想是使用对象、类、继承、封装、消息等基本概念来进行程序设计。本节主要介绍面向对象的三大基本特性：封装性、继承性与多态性。

2.2.1　类的封装

封装（Encapsulation）是将对象的属性和操作（或服务）结合为一个独立的整体的过程，封装是面向对象方法的重要原则。在 Java 中一般通过关键字 private 实现封装，将类的数据隐藏起来，控制用户对类的修改和访问数据的程度，是一种信息隐藏技术。

例 2-9　类的封装。

```
public class Show
{
    public static void show(String str)
    {
    System.out.println(str);
    }
    } //实现对 System.out.println()的封装
public class Use
{
    public static void main(String[] args)
    {
    Show.show("封装");
    }
    } //调用时不必使用 System.out.println("封装");
```

2.2.2 类的继承

对于 OOP 的程序来说，其精华就在于类的继承。继承是面向对象编程中的重要内容，允许创建分等级层次的类。运用继承，能够创建一个通用类，它定义了一系列相关项目的一般特性。该类可以被更具体的类继承，每个具体的类都增加一些自己特有的东西。被继承的类叫超类（父类），继承超类的类叫子类。

在 Java 中，子类对父类的继承是通过在类的定义过程中用关键字 extends 来说明的。

声明一个继承超类的类的通常形式如下：

```
class  子类名称 extends  父类名称
{
      //类的成员
}
```

继承一个类，只要用 extends 关键字把一个类的定义合并到另一个中就可以了，注意在 Java 中是单继承关系。

简单的一个范例如下：

```
class circle{  //父类
   private double pi =3.14;
   double radius ;
   其他方法…
}
class coin extends circle { //子类，即由circle类派生的类
   private int value; //新加入成员
   其他方法… //新加入方法
}
```

下面的例子创建了一个超类 A 和一个名为 B 的子类。注意怎样用关键字 extends 来创建 A 的一个子类。

例 2-10 使用关键字 extends 创建子类。

```
class A
{
        int i;
        int j;
        void showij()
        {
            System.out.println("i and j: " + i + " " + j);
        }
}
class B extends A
{
        int k;
        void showk()
        {
            System.out.println("k: " + k);
        }
        void sum()
        {
            System.out.println("i+j+k: " + (i+j+k));
}
```

```
}
class Test
{
        public static void main(String args[])
        {
                A a = new A();
                B b = new B();
                a.i = 10;
                a.j = 20;
                System.out.println("Contents of a: ");
                a.showij();
                System.out.println();
                b.i = 7;
                b.j = 8;
                b.k = 9;
                System.out.println("Contents of b: ");
                b.showij();
                b.showk();
                System.out.println();
                System.out.println("Sum of i, j and k in b:");
                b.sum();
        }
}
```

该程序的输出如下：

```
Contents of a:
i and j: 10 20
Contents of b:
i and j: 7 8
k: 9
Sum of i, j and k in b:
i+j+k: 24
```

在上例中，子类 B 包括它的超类 A 中的所有成员。子类 B 可以获取 i 和 j，以及调用 showij()方法。同样，sum()内部，i 和j 可以被 B 直接引用。

关于继承的几个规则：

（1）每个子类只能定义一个超类，Java 不支持多超类的继承；

（2）没有类可以成为它自己的超类；

（3）一个子类可以是另一个类的超类；

（4）子类不能访问超类中被声明成 private 的成员。

2.2.3 多态与方法重写

多态是指用相同的名字来定义不同的方法，即“一个接口，多个方法”。多态是面向对象编程的一大特征，体现了程序的可扩展性，也体现了程序代码的重复使用特性。2.1.5 小节中介绍的重载是类中多态性的一种表现；同时，方法重写是父类和子类多态性的一种表现。

方法重写中，子类方法的命名和父类中方法的命名是一致的，包括参数及返回类型一致。

例 2-11 方法重写。

```
class A
{
    void methodtest()
    {
```

```
            System.out.println("Inside A's methodtest method");
        }
    }
    class B extends A
    {
        void methodtest () //子类的方法和父类的方法相同
        {
            System.out.println("Inside B's methodtest method");
        }
    }
    class C extends A
    {
        void methodtest ()//子类的方法和父类的方法相同
        {
            System.out.println("Inside C's methodtest method");
        }
    }
    class Dispatch
    {
        public static void main(String args[])
        {
            A a = new A();
            B b = new B();
            C c = new C();
            A r;
            r = a;
            r. methodtest ();
            r = b;
            r. methodtest ();
            r = c;
            r. methodtest ();
        }
    }
```

该程序的输出如下：

```
Inside A's methodtest method
Inside B's methodtest method
Inside C's methodtest method
```

2.2.4 关键字 super

在 Java 语言中，用关键字 super 来指明对父类的引用。super 有两种通用形式。第 1 种调用超类的构造方法，用下面的形式：

super(parameter-list)。

这里 parameter-list 定义了超类中构造方法所用到的所有参数。第 2 种用来访问被子类的成员隐藏的超类成员。

例 2-12 执行父类的构造方法。

```
class subBox extends Box
{
    double weight;
    subBox (double w, double h, double d, double m)
    {
```

```
        super(w,h,d);  //执行父类的构造方法
        this.weight = m;
    }
}
```

例 2-13　使用super 调用超类的方法和变量。

```
class A
{
    int i;
    void printi()
    {
      System.out.println("A 中 i :"+i);
    }
}
class B extends A
{
    int i;
     B(int a, int b)
     {
        super.i = a;
        i = b;
     }
     void printi()
     {
        System.out.println("B 中 i :"+i);
     }
     void show()
     {
        System.out.println("i in superclass: " + super.i);
        System.out.println("i in subclass: " + i);
        super. printi();
        printi();
     }
}
class Test
{
     public static void main(String args[])
     {
        B b = new B(1, 2);
        b.show();
     }
}
```

该程序输出如下：

```
i in superclass: 1
i in subclass: 2
A 中 i :1
B 中 i :2
```

2.2.5　关键字 static

在 Java 语言中，通过关键字 static 修饰成员变量和方法，这时，它的使用完全独立于该类的任何对象，即不需要实例化就可以应用。通常情况下，类成员必须通过它的类的对象访问，但通过关键字 static 修饰就不必引用特定的实例。声明为 static 的变量实质上就是全局变量。

声明为 static 的方法有以下几条限制：它们仅能调用其他的 static 方法；它们只能访问 static 数据；它们不能以任何方式引用 this 或 super。

在 Java 中还可以声明一个 static 块，static 块仅在该类被加载时执行一次。StaticTest 类有一个 static 方法，一些 static 变量，以及一个 static 初始化块。

例 2-14　static 块使用实例。

```
class StaticTest
{
   static int a = 2;
   static int b;
   static void meth(int x)
   {
       System.out.println("x=" + x);
       System.out.println("a = " + a);
       System.out.println("b = " + b);
   }
   static
   {
       System.out.println("Static block initialized.");
       b = a * 5;
   }
public static void main(String args[])
{
       meth(10);
   }
}
```

该程序的输出如下：

```
Static block initialized.
x = 10
a = 2
b = 10
```

例 2-15　通过类名直接访问。

```
class StaticDemo
{
    static int a = 2;
    static int b = 9;
    static void callme()
    {
        System.out.println("a = " + a);
    }
}
class StaticByName
{
    public static void main(String args[])
    {
        StaticDemo.callme();
        System.out.println("b = " + StaticDemo.b);
    }
}
```

该程序的输出如下：

```
a = 2
b = 9
```

例 2-16　用 static 修饰的变量为一全局变量。

```
class test {
    static int i;
    int j;
    int getI(){
      return i;
    }
    int getJ(){
      return j;
    }
     public static void main(String args[])
     {
            test  t1 =  new test();
            test  t2 =  new test();
           t1.i = 10 ;
           t2.i = 20; //这时，t1.i 也为 20
           //常用 test.i 来表示，故也称为类变量
           t1.j =100;
           t2.j =200; //t1.j 为 100
           //实例变量
     }
}
```

2.2.6　final 类和 abstract 类

1. final 类

final 关键字修饰成员变量时用来定义一个常量，修饰方法时可以阻止重写 。final 关键字修饰的类称为 final 类，final 类没有子类且不能被继承。

例 2-17　修饰成员变量。

```
class A
{
     final double PI=3.14;
     void meth()
     {
          PI = 5;   //非法，不能被改变，是常量
          System.out.println("This is Illegal.");
     }
}
```

例 2-18　修饰方法。

```
class A
{
          final void meth()
          {
               System.out.println("This is a final method.");
          }
}
class B extends A
{
          void meth()   //出错，不能被重写
          {
               System.out.println("Illegal!");
          }
}
```

例 2-19　修饰类。

```
final class A
{
        // ...
}
class B extends A
{
        // ...
}
```

B 继承A 是不合法的，因为A 声明成final。

声明一个final 类则隐含说明它的所有方法也都是final。

2. abstract 类

关键字 abstract 修饰的类，称为 abstract（抽象）类。abstract 类只能声明对象，不能创建对象，不能通过new 操作符直接实例化。其只能定义一个被它的所有子类共享的通用形式，由每个子类自己去填写细节。声明一个抽象类，只需在类声明开始时在关键字 class 前使用关键字 abstract，通用形式如下：

```
[修饰符] abstract class classname extends parent implements interface ;
```

通过指定 abstract 类型修饰方法。这些方法被子类使用并实现，因为它们没有在超类中实现。声明一个抽象方法，用下面的通用形式：

```
[修饰符] abstract  返回类型 方法名称(参数列表);
```

任何含有一个或多个抽象方法的类都必须声明成抽象类；当不知道如何实现方法时，应将该类定义成一个抽象类，该方法定义成一个抽象方法。

例 2-20　abstract 类。

```
abstract class A
{
        abstract void abmeth();            //抽出方法，没有方法体
        void concrete ()
        {
            System.out.println("This is a concrete method.");
        }
}
class B extends A
{
        void abmeth ()                     //子类提供实现
        {
            System.out.println("B's implementation of callme.");
        }
}
class Test
{
        public static void main(String args[])
        {
            B b = new B();
            b. abmeth ();
            b. concrete ();
        }
}
```

2.2.7 类的接口

1. 接口的定义

接口是 Java 提供的一个重要功能，结构与抽象类非常相似，但是它们缺少实例变量，而且它们声明的方法是不含方法体的，方法体在实现接口的类中给出。一旦接口被定义，任何类成员都可以实现一个接口。接口表明了所有实现该接口类的共有属性和行为，只不过是每个类具体实现的方法内容不同。一个类只能有一个父类，却可以实现多个接口。

如果一个类实现一个接口，则必须实现接口中定义的所有方法。然而，每个类都可以自由地决定接口中方法实现的细节。声明接口使用关键字 interface，实现接口使用关键字 implements。

声明接口的通用形式：

```
[public] interface 接口名称
{
    [public] final 类型 常量名称=值;
    [public] 返回类型 方法名称(参数列表);
}
```

将类定义成 public，则处于任何位置的类都可以访问该接口，如果不指名接口的访问类型，则只有相同包中的类可以访问该接口的定义。

注意，接口中没有方法体，在默认的状态下接口中的所有属性都是 public static，所有方法都是 public；接口中不能定义静态方法，每个包含接口的类必须实现所有的方法。

接口声明中还可以声明变量。它们一般是 final 和 static 型的，它们的值不能通过实现类而改变，它们还必须以常量值初始化。如果接口本身定义成 public，所有方法和变量都是 public 的。

下面是一个接口定义的例子。它声明了一个简单的接口，该接口包含一个带单个整型参数的 callback()方法和一个 callbackto()方法。

```
interface Myinterface
{
        void callback(int param); //抽象方法
        void callbackto(); //抽象方法
}
```

另一个例子：

```
interface shape2D
{
   double pi = 3.14 ; //数据成员一定要初始化
   void area();//抽象方法
}
```

一旦接口被定义，一个或多个类都可以实现该接口。先用 implements 子句实现一个接口，然后创建接口定义的方法。一个包括 implements 子句的类的一般形式如下：

```
[修饰符] class 类名 [extends 父类名] [implements 接口 1 [,接口 2……]]
{
    // class-body
}
```

如果一个类实现一个接口，应对该接口中的方法提供全部的实现。如果不完全实现接口中的方法，则该类必须被定义成抽象类。如果一个类实现多个接口，那么这些接口被逗号分隔。如果一个类实现两个声明了同样方法的接口，那么相同的方法将被其中任意一个接口客户使用。实现接口的方法必须声明成 public。而且，创建方法体的名称、返回类型和参量必须严格与接口中声明的完全一致。

例 2-21 实现接口。

```
class Test implements Myinterface
{
        public void callback(int p)
        {
            System.out.println("test:"+ p);
        }
        public void callbackto()
        {
            System.out.println("test:");
        }
        public void other()
        {
            System.out.println("other:");
        }
}
```

下面的例子通过接口引用变量调用 callback()方法：

```
class  TestIface
{
    public  static  void  main(String  args[])
    {
        Myinterface  c  =  new  Test ();
        c.callback(10);
    }
}
```

该程序的输出如下：

```
test: 10
```

变量 c 被定义成接口类型 Myinterface，而且被一个 Test 实例赋值。尽管变量 c 可以用来访问 Callback()方法，但它不能访问 Test 类中的任何其他成员。一个接口引用变量仅仅知道被它的接口定义声明的方法。因此，变量 c 不能用来访问 other ()，因为它是被 Test 定义的，而不是由 Myinterface 定义。

如果一个类实现一个接口但是不完全实现接口定义的方法，那么该类必须定义成 abstract 型。

例 2-22 abstract 型定义类。

```
abstract  class  otherTest  implements  Myinterface
{
    int  a,  b;
    void show()
    {
        System.out.println(a + " " + b);
    }
    // ...
}
```

这里，类 otherTest 没有实现 callback()及 callbackto()方法，必须定义成抽象类。

接口与抽象类有两点不同：接口的数据成员必须初始化；接口里的方法必须全部声明成 abstract，即必须都是抽象方法。

2. 扩展接口

接口可以通过运用关键字 extends 被其他接口继承，其语法与继承类是一样的。原来的接口称为父接口，派生的接口称为子接口。当一个类实现一个继承了另一个接口的接口时，它必须实现接口继承链表中定义的所有方法。

接口扩展的语法为：

```
interface 子接口名字 extends 父接口名称 1，父接口名称 2,……
```

例 2-23　接口扩展。

```
interface A
{
    void a1();
    void a2();
}
interface B extends A
{
    void b1();
}
class MyClass implements B
{
    public void a1()
    {
        System.out.println("Implement a1().");
    }
    public void a2()
    {
        System.out.println("Implement a2().");
    }
    public void b1()
    {
        System.out.println("Implement b1().");
    }
}
class Test
{
    public static void main(String arg[])
    {
        MyClass ob = new MyClass();
        ob.a1();
        ob.a2();
        ob.b1();
    }
}
```

2.3　包

在 Java 中每个类都有自己的命名空间。Java 提供了把类名空间划分为更多易管理的块的机制，这种机制就是包。本节主要介绍包的概念、引入包的方法和包的编译。

2.3.1 包的概念

在 Java 中，为了方便管理，常常将类名空间划分为多个块，这种块的机制就是包，包在某种意义上说就是一个目录。包既是命名机制也是可见度控制机制。在 Java 中，既可以在包内定义类，而且在包外的代码不能访问该类，这样可使类相互之间有隐私。在编程时，将一个 package 命令作为一个 Java 源文件的第一句就可以定义包了。该文件中定义的任何类将属于指定的包。

包声明的通用形式：

```
package 包名;
```

例如，下面的声明创建了一个名为 myp 的包：

```
package myp;
```

Java 用文件系统目录来存储包。例如，声明的 myp 包中的类文件被存储在一个 myp 目录中，目录名必须和包名严格匹配。多个文件可以包含相同的 package 声明。包中可以含有子包，用点号"."分隔。一个多级包的声明的通用形式如下：

```
package pkg1[.pkg2[.pkg3]];
```

例 2-24 包的声明。

```
package test;
class Student
{
    public static void main(String args[])
   {
        System.out.println("This is……");
    }
}
```

文件名为 Student.java，把它存放在 test 目录中。编译文件，确信结果文件 Student.class 同样在 test 目录中。然后用下面的命令行执行：

```
java test. Student
```

执行该命令时必须在 test 的上级目录，或者把类路径环境变量设置成合适的值。

如上所述，Student 现在是 test 包的一部分，它不能自己执行，即不能用下面的命令行：

```
java Student
```

Student 必须和它的包名一起使用。

2.3.2 引入包

Java 包含了import 语句来引入特定的类甚至是整个包。import 语句体现了软件重用的思想。

在 Java 源程序文件中，import 语句紧接着 package 语句，它存在于任何类定义之前，下面是 import 声明的通用形式：

```
import pkg1[.pkg2].(classname|*);
```

这里，pkg1 是顶层包名，pkg2 是在外部包中的用逗点(.)隔离的下级包名。可以指定一个星号（*）引入整个包。下面的代码段显示了所用的两种形式：

```
import java.util.Date;
import java.io.*;
```

所有 Java 包含的标准 Java 类都存储在名为 java 的包中。基本语言功能被存储在 java 包中的 java.lang 包中。通常，Java 编译器为所有程序隐式地引入 java.lang 包。有时可以没有 import 语句，但也可以引入类。

例 2-25　有import 语句示例如下。

```
import java.util.*;
class MyDate extends Date
{
    ...
}
```

没有import 语句的例子如下：

```
class MyDate extends java.util.Date
{
    ...
}
```

2.3.3　访问保护

类和包都是封装和容纳名称空间和变量及方法范围的方法，Java 将类成员的可见度分为以下 4 个种类：

（1）相同包中的子类；

（2）相同包中的非子类；

（3）不同包中的子类；

（4）既不在相同包，又不在相同子类中的类。

3 个访问控制符 private、public 和 protected，提供了多种方法来产生这些种类所需访问的多个级别，表 2.1 总结了它们之间的相互作用。

表 2.1　访问控制

	Private 成员	默认的成员	Protected 成员	Public 成员
同一类中可见	√	√	√	√
同一个包中对子类可见		√	√	√
同一个包中对非子类可见		√	√	√
不同包中对子类可见			√	√
不同的包中对非子类可见				√

2.3.4　包的编译

在 Java 源文件代码中，如果含有 package 语句，则编译时应指定编译后的包放在哪个目录下。只要在编译时带一个参数即可，例如：

javac –d 包存放目录文件名.java

在“-d”后面带一个点号代表当前目录。

如希望将所有包都存放在一个集中目录“c:\mypack”中，则可以先将这个目录创建好，再进行如下编译：

```
javac -d c:\mypack mydemo.java
```

思考与习题

1. 填空题：面向对象具有________、________、________3 个特性。

2. 填空题：封装（Encapsulation）是________________的过程，封装是面向对象方法的重要原则。

3. 填空题：多态是________________的方法，即“一个接口，多个方法”。

4. 简答题：简述类、对象的概念及其区别。

5. 简答题：简述 final 关键字和 static 关键字的修饰类、方法和属性的含义。

6. 简答题：简述包的含义，并举例说明。

7. 编程题：对学生的成绩进行管理，即设计学生类、课程类、老师类，要求有继承类、重载和重写方法。

8. 编程题：定义一个复数类，该数具有整数类型的两个属性 a 和 b，分别代表一个复数的实部和虚部，对应于形式如 a+bi 形式的复数形式。该类具有多个构造方法、复数加、减、乘、除运算，并输出结果。

第3章 线程和网络编程

本章介绍线程和网络编程的相关知识。线程是资源分配和调度的基本单位，本章主要从线程的概念、Java 线程的模型、创建线程和线程同步等方面介绍线程。网络编程方面重点介绍了基于 TCP/UDP 的编程，并通过实例具体说明。

3.1 线程概念

在操作系统中都会涉及处理机管理，即会引入一个进程的概念，它是动态的、相互隔离的、独立运行的，是资源分配单位及调度的基本单位。进程是由进程控制块、程序段、数据段 3 部分组成的。线程是共存于一个进程中的多个并发执行流。大多数现代的操作系统都支持线程，而且线程的概念已经以各种形式存在。Java 是第一个在语言本身中显式地包含线程的主流编程语言，而不是对底层操作系统的调用。Java 支持多线程编程（multithreaded programming）。多线程程序包含两条或两条以上并发执行的部分，程序中每个部分都叫一个线程（thread），每个线程都有独立的执行路径。多任务处理有两种类型：基于进程的和基于线程的。多线程是多任务处理的一种特殊形式。

进程与线程是有区别的。进程（process）本质上是一个并发执行的程序。因此，基于进程（process-based）的多任务处理的特点是允许计算机同时运行两个或更多的程序。基于线程（thread-based）的多任务处理环境中，线程是最小的执行单位，单线程具有自己的堆栈、程序计数器和局部变量。进程是重量级的任务，需要分配自己独立的地址空间。进程间的通信是昂贵和受限的，进程切换也需要花费大量的处理机时间。而线程是轻量级的任务，共享相同的地址空间，并且共同分享同一个进程，线程间的通信和切换是低成本的。

3.1.1 Java 线程模型

Java 运行系统在很多方面依赖于线程，所有的类库设计都考虑到多线程。实际上，Java 使用线程来使整个环境异步。这有利于通过防止 CPU 循环的浪费来减少无效部分。

在单线程处理环境中，是使用一种叫作轮询的事件循环方法。在该模型中，单线程控制在无限循环中运行，轮询一个事件序列来决定下一步做什么，当一个事件在处理时，其他事件必须等待，这就浪费了 CPU 时间。这将导致程序的一部分独占系统，阻止了其他事件的执行。总体来说，单线程环境下，当一个线程因为等待资源时阻塞，整个程序停止运行。

Java 采用多线程处理模型，其优点在于取消了主循环/轮询机制。一个线程的暂停并不影响程序的其他部分，即多个线程通过系统调度并发执行。若 Java 程序中出现线程阻塞，仅有一个线程暂停，其他线程继续运行。

每个线程都有产生、运行、消亡的状态转换过程，线程的状态如图 3.1 所示。

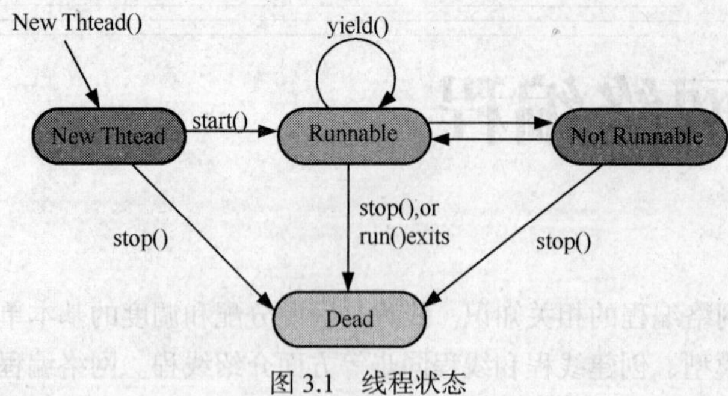

图 3.1　线程状态

创建状态（new thread）：执行“Thread myThread = new MyThreadClass( );”语句时，线程就处于创建状态。当一个线程处于创建状态时，它仅仅是一个空的线程对象，系统不为它分配资源，尚未执行（start()尚未调用）。

可运行状态（runnable）：当一个线程处于可运行状态时，系统为这个线程分配了它需要的系统资源，安排其运行并调用线程运行方法，这样就使得该线程处于可运行（runnable）状态。

如执行：Thread myThread = new MyThreadClass( );

```
myThread.start( );
```

需要注意的是这一状态并不是运行中状态（running），因为线程也许实际上并未真正运行。由于很多计算机都是单处理器的，所以要在同一时刻运行所有处于可运行状态的线程是不可能的，Java 的运行系统必须实现调度来保证这些线程共享处理器。

不可运行状态（not runnable）：不可运行状态也称为阻塞状态（Blocked）。因为某种原因（输入/输出、等待消息或其他阻塞情况），系统不能执行线程的状态，这时即使处理器空闲，也不能执行该线程。进入不可运行状态的原因如下：

（1）调用了 sleep()方法；

（2）调用了 suspend()方法；

（3）为等候一个条件变量，线程调用 wait()方法；

（4）输入/输出流中发生线程阻塞。

死亡状态（dead）：线程的终止一般可通过两种方法实现，即自然撤销（线程执行完）或是被停止（调用 stop()方法）。不推荐通过调用 stop()来终止线程的执行，而是让线程执行完。

Java 给每个线程安排优先级以决定其执行顺序。线程优先级是详细说明线程间优先关系的数值。线程可以被高优先级的线程抢占，这叫作有优先权的多任务处理，即系统中运行的线程一定是高优先级的。低优先级的线程被高优先级的线程剥夺处理机执行权。

多线程在程序中引入了一个异步行为，需要时必须有加强同步性的方法。举例来说，如果希望两个线程相互通信并共享一个队列时，就需要某些方法来确保它们没有相互冲突。必须防止一个线程正在入队列而另一个线程正在出队列。

把程序分成若干线程后，就要定义各线程之间的联系。在用大多数其他语言规划时，你必须依赖于操作系统来确立线程间通信。Java 的消息传递系统允许一个线程进入一个对象的一个同步方法，然后调用 Wait()方法进行等待，直到其他线程通过 Notify()方法明确通知它出来。

Java 的多线程系统建立于 Thread 类。Thread 类封装了线程的执行。为创建一个新的线程，程序必须扩展 Thread 或实现 Runnable 接口。

Thread 类定义了多种方法来帮助管理线程。本章用到的方法如表 3.1 所示。

表 3.1　Thread 类方法

方　法	意　义
getName	获得线程名称
getPriority	获得线程优先级
isAlive	判定线程是否仍在运行
join	等待一个线程终止
run	线程的入口点
sleep	在一段时间内挂起线程
start	通过调用运行方法来启动线程

3.1.2　主线程

Java 程序至少包含一个线程，即当 Java 程序启动时，一个线程立刻运行，该线程通常叫作程序的主线程（main thread），因为它是程序开始时就执行的。主线程的重要性体现在两方面：它是产生其他子线程的线程；通常它必须最后完成执行，因为它执行各种关闭动作。

尽管主线程在程序启动时自动创建，但它可以由一个 Thread 对象控制。为此，必须调用方法 currentThread()获得它的一个引用，currentThread()是 Thread 类的公有的静态成员。它的通常形式如下：

```
static Thread currentThread()
```

该方法返回一个调用它的线程的引用。一旦获得主线程的引用，就可以像控制其他线程那样控制主线程。

例 3-1　调用 currentThread()获得当前线程的引用。

```
public class Test
{
    public  static  void  main(String  args[])
    {
        Thread ct= Thread.currentThread();
        System.out.println("Current thread: " + ct);
        ct.setName("Test");
        System.out.println("After name change: " + ct);
        try
        {
            Thread.sleep(1000);
        }
        catch (InterruptedException e)
        {
                System.out.println("Main thread interrupted");
        }
    }
}
```

程序运行结果：

```
Current thread: Thread[main,5,main]
After name change: Thread[Test,5,main]
```

在本程序中，当前线程（自然是主线程）的引用通过调用 currentThread()获得，该引用保存在局部变量 ct 中。然后，程序显示了线程的信息。接着程序调用 setName()改变线程的内部名称，线程信息又被显示。Sleep()语句明确规定延迟时间是 1ms。注意循环外的 try/catch 块。Thread 类的 Sleep()方法可能引发一个 InterruptedException 异常。

println()中输出显示顺序：线程名称、优先级、组的名称。默认情况下，主线程的名称是 main。它的优先级是 5，这也是默认值，main 也是所属线程组的名称。一个线程组是一种将线程作为一个整体集合的状态控制的数据结构。

3.1.3 创建线程

大多数情况，通过实例化一个 Thread 对象来创建一个线程。Java 定义了两种方式：实现 Runnable 接口和继承 Thread 类。

1. 实现 Runnable 接口

创建线程最简单的方法就是创建一个实现 Runnable 接口的类。Runnable 抽象了一个执行代码单元，可以通过实现 Runnable 接口的方法创建每一个对象的线程。为实现 Runnable 接口，一个类仅需实现一个 run()的简单方法，该方法声明如下：

```
public void run()
```

在 run()中可以定义代码来构建新的线程。run()方法能够像主线程那样调用其他方法，引用其他类，声明变量。仅有的不同是 run()在程序中确立另一个并发的线程执行入口。当 run()返回时，该线程结束。

在已经创建了实现Runnable接口的类以后，需要在类内部实例化一个Thread类的对象。Thread类定义了多种构造方法，例如：

```
Thread(Runnable threadOb, String threadName)
```

该构造方法中，threadOb 是一个实现 Runnable 接口类的实例，这定义了线程执行的起点。新线程的名称由 threadName 定义。建立新的线程后，它并不运行直到调用了它的 start()方法，该方法在 Thread 类中定义。本质上，start()执行的是一个对 run()的调用。start()方法声明如下：

```
void start()
```

例 3-2 创建线程。

```
class Childt  implements  Runnable
{
     Thread  t;
     Public void NewThread()
    {
       t = new Thread(this, " Childt Thread");
       System.out.println("Child thread: " + t);
       t.start();
       }
     public void run()
     {
       try
```

```
    {
        for(int i = 5; i > 0; i--)
        {
            System.out.println("Child Thread: " + i);
            Thread.sleep(500);
        }
    }
    catch (InterruptedException e)
    {
      System.out.println("Child interrupted.");
    }
    System.out.println("Exiting child thread.");
  }
}
public class  Test
{
    public static void main(String args[])
    {
        Childt c=new Childt();
        c.new NewThread();
        try
        {
            for(int i = 5; i > 0; i--)
            {
                System.out.println("Main Thread: " + i);
                Thread.sleep(1000);
            }
        }
        catch (InterruptedException e)
        {
            System.out.println("Main thread interrupted.");
        }
        System.out.println("Main thread exiting.");
    }
}
```

该程序的输出如下：

```
Child  thread:  Thread[Childt Thread,5,main]
Main  Thread:  5
Child  Thread:  5
Child  Thread:  4
Childt Thread: 3
Main  Thread:  4
Child  Thread:  2
Main  Thread:  3
Child  Thread:  1
Exiting  child  thread.
Main  Thread:  2
Main  Thread:  1
Main  thread  exiting.
```

在 Childt 构造方法中，新的 Thread 对象由下面的语句创建：

```
t = new Thread(this, " Childt Thread ");
```

this 指代当前新创建的线程，然后调用 start()，执行体在 run()方法中，这使子线程 for 循环开始执行。调用 start()之后，Childt 的构造方法返回到 main()。当主线程被恢复时，它到达 for 循环。

两个线程继续运行，共享 CPU，直到它们的循环结束。如前面提到的，在多线程程序中，通常主线程必须是结束运行的最后一个线程。上述程序保证了主线程最后结束，因为主线程沉睡周期为 1 000ms，而子线程仅为 500ms，这就使子线程在主线程结束之前先结束。

2. 扩展 Thread

创建线程的另一个途径是创建一个新类来扩展 Thread 类，然后创建该类的实例。当一个类继承 Thread 时，它必须重载 run()方法，这个 run()方法是新线程的入口。它也必须调用 start()方法去启动新线程执行。下面用扩展 thread 类重写前面的程序。

例 3-3 扩展 Thread 类。

```
class Mythread extends Thread
{
     Mythread ()
     {
         super("Demo Thread");
         System.out.println("Child thread: " + this);
         start();
     }
     public void run()
     {
       try
       {
         for(int i = 5; i > 0; i--)
         {
              System.out.println("Child Thread: " + i);
              Thread.sleep(500);
         }
       }
       catch (InterruptedException e)
       {
         System.out.println("Child interrupted.");
       }
       System.out.println("Exiting child thread.");
     }
}
public class Test
{
  public static void main(String args[])
  {
     new Mythread ();
     try
     {
         for(int i = 5; i > 0; i--)
         {
              System.out.println("Main Thread: " + i);
              Thread.sleep(1000);
         }
     }
     catch (InterruptedException e)
     {
         System.out.println("Main thread interrupted.");
         }
     System.out.println("Main thread exiting.");
  }
}
```

该程序生成和前面版本相同的输出。子线程是由实例化 Mythread 对象生成的，该对象从 Thread 类派生。

Java 提供了两种创建子线程的方法，当已经继承了一个父类或仅需要重载 run()方法时，可以使用 Runnable 接口，否则都可以使用。

到目前为止，我们仅用到两个线程：主线程和一个子线程。然而，程序可以创建所需的更多线程。

例 3-4　下面的程序创建了 3 个子线程。

```
class  NewThread  implements  Runnable
{
     String  name;
     Thread  t;
     NewThread(String  threadname)
     {
       name = threadname;
       t = new Thread(this, name);
       System.out.println("New thread: " + t);
       t.start();
     }
     public void run()
     {
       try
       {
          for(int i = 5; i > 0; i--)
          {
                System.out.println(name + ": " + i);
                Thread.sleep(1000);
          }
        }
        catch (InterruptedException e)
        {
          System.out.println(name + "Interrupted");
        }
        System.out.println(name + " exiting.");
     }
}
public class Test
{
     public static void main(String args[])
     {
       new NewThread("One");
       new NewThread("Two");
       new NewThread("Three");
       try
       {
            Thread.sleep(10000);
       }
       catch (InterruptedException e)
       {
            System.out.println("Main thread Interrupted");
       }
       System.out.println("Main thread exiting.");
     }
}
```

程序输出如下：

```
New thread: Thread[One,5,main]
New thread: Thread[Two,5,main]
New thread: Thread[Three,5,main]
One: 5
Two: 5
Three: 5
One: 4
Two: 4
Three: 4
One: 3
Three: 3
Two: 3
One: 2
Three: 2
Two: 2
One: 1
Three: 1
Two: 1
One exiting.
Two exiting.
Three exiting.
Main thread exiting.
```

程序一旦启动，所有 3 个子线程共享 CPU。注意 main()中对 sleep(10000)的调用。这使主线程沉睡 10s 以确保它最后结束。

在 Java 中，有两种方法可以判定一个线程是否结束。其中一种方法是在线程中调用 isAlive()。这种方法由 Thread 定义，它的通常形式如下：

```
final boolean isAlive();
```

如果所调用线程仍在运行，isAlive()方法返回 true：如果不是，则返回 false。但 isAlive()很少用到，等待线程结束的更常用的方法是调用 join()，描述如下：

```
final void join() throws InterruptedException;
```

该方法等待所调用线程结束。

3.1.4 线程同步

当两个或两个以上的线程需要共享资源，此时需要某种方法来确定资源在某一刻仅被一个线程占用，达到此目的的过程叫作同步（synchronization）。在某些情况中，没有必要用同步来将数据从一个线程传递到另一个，因为 JVM 已经隐含地执行同步了，如由静态初始化器（在静态字段上或 static{}块中的初始化器）初始化数据时、访问 final 字段时，以及在创建线程之前创建对象时等。

如果没有同步，数据很容易就处于不一致状态。例如，如果一个线程正在更新两个相关值（如粒子的位置和速率），而另一个线程正在读取这两个值，有可能在第 1 个线程只写了一个值，还没有写另一个值的时候，调度第 2 个线程运行，这样它就会看到一个旧值和一个新值。同步让我们可以定义必须同时运行的代码块，这样对于其他线程而言，它们要么都执行，要么都不执行。

同步的关键是管程（也叫信号量 semaphore），管程是一个互斥独占锁定的对象，或称互斥体（mutex）。在给定的时间，仅有一个线程可以获得管程。当一个线程需要锁定时，它必须进入管程，此时其他试图进入锁定管程的线程必须挂起直到第 1 个线程退出。这些其他的线程被称为等待管程。一个拥有管程的线程可以再次进入相同的管程。

Java 中同步是简单的，因为所有对象都有与它们对应的隐式管程。通过调用被 synchronized 关键字修饰的方法进入某一特定对象的管程。当一个线程在一个同步方法内部，所有试图调用该方法（或其他同步方法）的同实例的其他线程必须等待。为了退出管程，并将对象的控制权转交给其他等待的线程，拥有管程的线程仅需从同步方法中返回。

例 3-5　线程同步。

```
class Callme
{
   void call(String msg)
   {
    System.out.print("[" + msg"]");
    try
    {
      Thread.sleep(1000);
    }
    catch(InterruptedException e)
    {
      System.out.println("Interrupted");
    }
    System.out.println(")");
   }
}
class Caller implements Runnable
{
   String msg;
   Callme target;
   Thread t;
   public Caller(Callme targ, String s)
   {
    target = targ;
    msg = s;
    t = new Thread(this);
    t.start();
   }
   public void run()
   {
    target.call(msg);
   }
}
public class Test
{
   public static void main(String args[])
   {
    Callme target = new Callme();
    Caller ob1 = new Caller(target, "Hello");
    Caller ob2 = new Caller(target, "Synchronized");
    Caller ob3 = new Caller(target, "World");
    try
    {
```

```
            ob1.t.join();
            ob2.t.join();
            ob3.t.join();
        }
        catch(InterruptedException e)
        {
            System.out.println("Interrupted");
        }
      }
    }
```

该程序的输出如下：

```
[Hello][Synchronized][World]
)
)
```

从输出看，结果并非想要的那样。因为通过调用 sleep()、call()方法允许执行转换到另一个线程，导致该结果是 3 个消息字符串的混合输出。如果能够限制 3 个线程同时调用同一对象的同一方法，则可以得到期望结果。

为达到上述目的，必须有权连续地使用 call()。也就是说，在某一时刻，必须限制只有一个线程可以支配它。为此，只需在 call()定义前加上关键字 synchronized，例如：

```
synchronized void call(String msg)
```

这样保证了在一个线程使用 call()时其他线程进入 call()。在 synchronized 加到 call()前面以后，程序输出如下：

```
[Hello]
[Synchronized]
[World]
```

在有些情况下，不能修改类代码（如该类是第三方提供的），那么如何来同步线程呢？解决方法很简单：只需将对这个类定义的方法的调用放入一个 synchronized 块内就可以了。

下面是 synchronized 语句的普通形式：

```
synchronized(object)
{
    //  statements  to  be  synchronized
}
```

其中，object 是被同步对象的引用。

下面是前面程序的修改版本，在 run()方法内用了同步块：

```
class  Callme
{
   void  call(String  msg)
   {
     System.out.print("[" + msg"]");
     try
     {
         Thread.sleep(1000);
     }
     catch (InterruptedException e)
     {
         System.out.println("Interrupted");
```

```
      }
      System.out.println(")");
    }
  }
  class Caller implements Runnable
  {
    String msg;
    Callme target;
    Thread t;
    public Caller(Callme targ, String s)
    {
      target = targ;
      msg = s;
      t = new Thread(this);
      t.start();
    }
    public void run()
    {
      synchronized(target)
      {
          target.call(msg);
      }
    }
  }
  public class Test
  {
    public static void main(String args[])
    {
      Callme target = new Callme();
      Caller ob1 = new Caller(target, "Hello");
      Caller ob2 = new Caller(target, "Synchronized");
      Caller ob3 = new Caller(target, "World");
      try
      {
          ob1.t.join();
          ob2.t.join();
          ob3.t.join();
      }
      catch(InterruptedException e)
      {
          System.out.println("Interrupted");
      }
    }
  }
```

这里 call()方法没有被 synchronized 修饰，而 synchronized 是在 Caller 类的 run()方法中声明的。这可以得到上例中同样正确的结果，因为每个线程运行前都等待先前的一个线程结束。

3.1.5　线程通信

前面的例子都无条件地阻塞了其他线程异步访问某个方法。多线程通过把任务分成离散的、合乎逻辑的单元代替事件循环程序。

Java 包含了通过 wait()、notify()和 notifyAll()方法实现的一个进程间通信机制。这些方法在对象中是用 final 方法实现的，所以所有的类都含有它们。这 3 种方法仅在 synchronized 方法中才能被调用。

wait()告知被调用的线程放弃管程进入睡眠直到其他线程进入相同管程并且调用 notify()。

notify()恢复相同对象中第 1 个调用 wait()的线程。

notifyAll()恢复相同对象中所有调用 wait()的线程，具有最高优先级的线程最先运行。

这些方法在 Object 中被声明，例如：

```
final void wait() throws InterruptedException
final void notify()
final void notifyAll()
```

例 3-6 程序实行了一个简单生产者/消费者的问题。它由 4 个类组成：Q（队列，模拟取数据、存数据）、Producer（生产者线程对象）、Consumer（消费者线程对象），以及主程序类。

例 3-6 简单生产者/消费者的问题实例。

```
class Q
{
   int n;
   boolean valueSet = false;
   synchronized int get()
   {
     if(!valueSet)
           try
           {
                wait();
           }
           catch(InterruptedException e)
           {
           System.out.println("InterruptedException caught");
           }
       System.out.println("Got: " + n);
       valueSet = false;
       notify();
       return n;
   }
   synchronized void put(int n)
   {
     if(valueSet)
           try
           {
                wait();
           }
           catch(InterruptedException e)
           {
           System.out.println("InterruptedException caught");
           }
     this.n = n;
     valueSet = true;
     System.out.println("Put: " + n);
     notify();
   }
}
class Producer implements Runnable
{
   Q q;
```

```
    Producer(Q q)
    {
      this.q = q;
      new Thread(this, "Producer").start();
    }
    public void run()
    {
      int i = 0;
      while(true)
      {
          q.put(i++);
      }
    }
}
class Consumer implements Runnable
{
    Q q;
    Consumer(Q q)
    {
      this.q = q;
      new Thread(this, "Consumer").start();
    }
    public void run()
    {
      while(true)
      {
          q.get();
      }
    }
}
public class Test
{
    public static void main(String args[])
    {
      Q q = new Q();
      new Producer(q);
      new Consumer(q);
      System.out.println("Press Control-C to stop.");
    }
}
```

下面是该程序的输出，它清楚地显示了同步行为：

```
Put: 1
Got: 1
Put: 2
Got: 2
Put: 3
Got: 3
Put: 4
Got: 4
Put: 5
Got: 5
```

3.2 网络编程基础

网络编程技术是当前一种主流的编程技术，随着互联网趋势的逐步增强以及网络应用程序的大量出现，网络编程技术在实际的开发中获得了大量的使用。网络编程的实质就是两个（或多个）设备（如计算机）之间的数据传输。本节中尽量以浅显的基础知识和实际的案例为读者介绍网络编程基础知识，如网络编程的基本概念和术语，以及基于 TCP/UDP 的编程实例。而在后续的实践中，读者仍需要大量实际的练习才能够在网络编程方面有所造诣。

3.2.1 TCP/UDP

1. TCP 概念

TCP（Transmission Control Protocol）是传输控制协议，它主要负责数据的分组和重组。在简化的计算机网络 OSI 模型中，它完成第 4 层传输层所指定的功能。在因特网协议族（Internet protocol suite）中，TCP 层是位于 IP 层之上、应用层之下的传输层。TCP 是一个“可靠的”、面向连接的传输机制，通过 3 次握手建立连接。它提供一种可靠的字节流保证数据完整、无损，并且按顺序到达。TCP 试图将数据按照规定的顺序发送，这在实时数据流或者路由高网络层丢失率应用的时候可能成为一个缺陷。TCP 提供服务的主要特点是：面向连接的传输；端到端的通信；高可靠性，确保传输数据的正确性，不出现丢失或乱序；全双工方式传输；采用字节流方式，即以字节为单位传输字节序列；紧急数据传送功能。

2. UDP 概念

UDP（User Datagram Protocol）是用户数据报协议，它和 TCP 一样都是网络传输层上的协议，但与 TCP 有本质的区别。UDP 是“不可靠”、无连接和面向消息的协议，它使用数据报进行传输。UDP 通信时不需要接收方确认，属于不可靠的传输，可能会出现丢包现象，并且不保证它们按顺序到达。但是 UDP 占用资源比较少，所以一般在一些可靠性要求比较低的网络上应用，如网络视频会议、在线影视和聊天等音频、视频数据传送。

3.2.2 端口

端口（Port）可以被理解成计算机与外界通信交流的窗户。网络上的一台计算机可以提供多种服务，单靠 IP 地址无法将它们区别开，所以通过“IP 地址+端口号”的形式来区分不同的服务。端口号是一个整数，通常范围在 0 ~ 65 535。小于 1024 的端口号一般分配给特定的服务协议，如：远程登录（Telnet）端口号为 23、电子邮件（SMTP）端口号为 25、文件传输（FTP）端口号为 21、超文本传输协议（Http）端口号为 80 等。如果是自定义的端口号，应尽量大于 1024。

3.2.3 套接字

套接字是支持 TCP/IP 的网络通信的基本操作单元，可以看成在两个程序进行通信连接中的一个端点，是连接应用程序和网络驱动程序的桥梁，Socket 在应用程序中创建，通过绑定与网络驱动建立关系。简单地说就是通信双方的一种约定，用套接字中的相关函数来完成通信过程。网络套接字（network socket）有一点像电源插座。网络周围的各式插头有一个标准方法传输它们的有效负载。理解标准协议的任何东西都能够插入套接字并进行通信。对于电源插座，不论插入一个电灯或是烤箱，只要它们使用 50Hz、220V 电压，设备将会工作。

常用的 TCP/IP 的 3 种套接字类型，即流套接字（SOCK_STREAM）、数据报套接字（SOCK_DGRAM）和原始套接字（SOCK_RAW）。

流套接字（SOCK_STREAM）：流套接字用于提供面向连接、可靠的数据传输服务。该服务将保证数据能够实现无差错、无重复发送，并按顺序接收。流套接字之所以能够实现可靠的数据服务，原因在于其使用了传输控制协议，即 TCP。

数据报套接字（SOCK_DGRAM）：数据报套接字提供了一种无连接的服务。该服务并不能保证数据传输的可靠性，数据有可能在传输过程中丢失或出现数据重复，且无法保证顺序地接收到数据。数据报套接字使用 UDP 进行数据的传输。

原始套接字（SOCK_RAW）：原始套接字可以读写内核没有处理的 IP 数据包，如果要访问其他协议，发送数据必须使用原始套接字。

3.2.4　客户机/服务器模式

客户机/服务器（Client/Server，C/S）模式又称 C/S 结构，是 20 世纪 80 年代末逐步发展起来的一种模式，是软件系统体系结构的一种。C/S 结构的关键在于功能的分布，一些功能放在前端机（即客户机）上执行，另一些功能放在后端机（即服务器）上执行。功能的分布在于减少计算机系统的各种瓶颈问题。简单地讲 C/S 模式就是基于企业内部网络的应用系统。

客户和服务器之间的连接就像电灯和电源插头的连接。房间的电源插座是服务器，电灯是客户。服务器是永久的资源，在访问过服务器之后，客户可以“拔去插头”。

在 Berkeley 套接字中，套接字的概念允许单个计算机同时服务于很多不同的客户，并能够提供不同类型信息的服务。该种技术由引入的端口处理，此端口是一个特定机器上的被编号的套接字。服务器进程是在“监听”端口直到客户连到它。尽管每个客户部分是独特的，一个服务器允许在同样端口接受多个客户。为管理多个客户连接，服务器进程必须是多线程的，或者有同步输入/输出处理多路复用技术的其他方法。

客户机/服务器的通信基于套接字。套接字是 internet 通信的端点，可以理解为是客户机和服务器之间的两端。客户程序创建客户套接字，服务器应用程序创建服务器套接字。双方的套接字连接起来后，数据通过这一连接来交换，如图 3.2 所示。

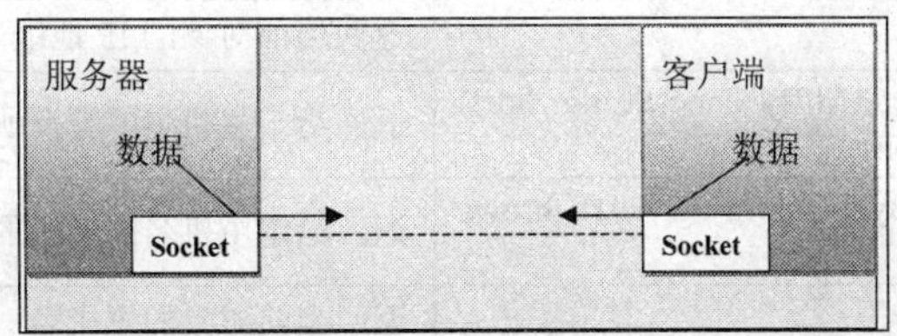

图 3.2　套接字连接

3.2.5　Java 和网络

Java 通过扩展前面介绍的已有的流式输入/输出接口和增加在网络上建立输入/输出对象特性这两种方法支持 TCP/IP。Java 支持 TCP 和 UDP 协议族。Java 提供的网络功能有 3 大类：URL、Socket 和 Datagram。TCP 用于网络可靠的流式输入/输出。UDP 支持更简单的、快速的、点对点的数据报模式。Java 用于网络操作的功能包是 java.net，它包含多个访问各种标准网络协议的类包。

表 3.2 所示为 java.net 所包含的类，表 3.3 所示为 java.net 包的接口。

表 3.2　网络类 java.net 包所包含的类

Authenticator (Java 2)	JarURLConnection (Java 2)	SocketPermission
ContentHandler	MulticastSocket	URL
DatagramPacket	NetPermission	URLClassLoader (Java 2)
DatagramSocket	PasswordAuthentication (Java 2)	URLConnection
DatagramSocketImpl	ServerSocket	URLDecoder (Java2)
HttpURLConnection	Socket	URLEncoder
InetAddress	SocketImpl	URLStreamHandler

表 3.3　java.net 包的接口

ContentHandlerFactory	SocketImplFactory	URLStreamHandlerFactory
FileNameMap	SocketOptions	DatagramSocketImplFactory

3.2.6　InetAddress 类

无论是打电话、发送邮件或建立与 Internet 的连接，地址都是基础。InetAddress 类用来封装前面讨论的数字式的 IP 地址和该地址的域名。通过 IP 主机名与这个类发生作用，IP 主机名比它的 IP 地址用起来更简便，更容易理解。InetAddress 类内部隐藏了地址数字，也就是说 InetAddress 类可以用于标识网络上的硬件资源，建立 IP 地址。把 IP 地址或是 Domain Name 转换成计算机看得懂的网络地址。

InetAddress 类没有明显的构造方法。为生成一个 InetAddress 对象，必须运用一个可用的工厂方法。工厂方法仅是一个类中静态方法返回一个该类实例的约定。这是在一个带有各种参数列表的重载构造方法中完成的，当持有唯一方法名时可使结果更清晰。InetAddress 类主要方法如表 3.4 所示。

表 3.4　InetAddress 类主要方法

方 法 名 称	方 法 说 明
public Boolean equals(Object obj)	判断给定对象是否与当前对象拥有相同的 IP 地址。相同时，函数返回 true；不同时，返回 false
public byte[] getAddress()	返回当前对象的 IP 地址
public static InetAddress[] getAllByName(String host) throws UnknownHostException	返回给定主机名的资源所在的所有 IP 地址
public static InetAddress getByName(String host) throws UnknownHostException	返回给定主机名的主机的 IP 地址
public String getHostName()	返回当前对象的主机名
public static InetAddress getLocalHost() throws Unknown HostException	返回本地主机的 IP 地址
public int hashCode()	返回当前对象的 IP 地址的散列码
public String toString()	返回当前对象的 IP 地址的字符串表示

例 3-7　InetAddress 类运用的实例。

```
import java.net.*;
public class Test
{
```

```
    public static void main(String args[]) throws UnknownHostException
    {
        InetAddress Address = InetAddress.getLocalHost();
        System.out.println(Address);
        Address = InetAddress.getByName("myweb.com");
        System.out.println(Address);
        InetAddress ia[] = InetAddress.getAllByName("www. myweb.com");
        for (int i=0; i< ia.length; i++)
           System.out.println(ia[i]);
    }
}
```

InetAddress 类也有一些非静态的方法，列于表 3.5 中，它们可以用于前面讨论过的方法返回的对象。

表 3.5　InetAddress 类方法

方　法	含　义
boolean equals(Object other)	如果对象具有和 other 相同的 Internet 地址则返回 true
byte[ ] getAddress()	返回代表对象的 Internet 地址的以网络字节为顺序的有 4 个元素的字节数组
String getHostAddress()	返回代表与 InetAddress 对象相关的主机地址的字符串
String getHostName()	返回代表与 InetAddress 对象相关的主机名的字符串
int hashCode()	返回调用对象的散列码
boolean isMulticastAddress()	如果 Internet 地址是一个多播地址，返回 true；否则，返回 false
String toString()	返回主机名字符串和 IP 地址

Internet 地址在分层的缓存服务器系列中被找到。这意味着本地机可能像知道它自己和附近的服务器一样知道一个名称——IP 地址的自动映射。对于其他名称，它可能向一个本地 DNS 服务器询问 IP 地址信息。如果那个服务器不含有一个指定的地址，它可以到一个远程的站点去询问。

3.2.7　URL

Web 是一个由 Web 浏览器统一的高级协议和文件格式的松散集合。URL 提供了一个容易理解的形式唯一确定或对 Internet 上的信息进行编址。URL 是无所不在的，每一个浏览器用它们来识别 Web 上的信息。在 Java 的网络类库中，URL 类为用 URL 在 Internet 上获取信息提供了一个简单的、简洁的用户编程接口（API）。

“http://www.myweb.com/” 和 “http://www. myweb.com:80/index.htm” 是 URL 的两个例子。一个 URL 规范以 4 个元素为基础。第 1 个元素是所用到的协议，用冒号（:）来将它与定位符的其他部分相隔离。第 2 个元素是主机名或所用主机的 IP 地址，这由左边的双斜线（//）和右边的单斜线（/）或可选冒号（:）限制。第 3 个元素是端口号，是可选的参数，由主机名左边的冒号（:）和右边的斜线（/）限制（它的默认端口为 80，它是预定义的 HTTP 端口，“:80” 是多余的）。第 4 个元素是实际的文件路径（或称应用程序）。多数 HTTP 服务器将给 URL 附加一个与目录资源相关的 index.html 或 index.htm 文件，所以，“http://www.myweb.com/” 与 “http://www. myweb.com/index.htm” 是相同的。

Java 的 URL 类的主要方法如表 3.6 所示。

表 3.6　URL 类的主要方法

方法名称	方法说明
public URL(String addr) throws MalformedURLException	创建一个给定资源地址的 URL 对象，这里，addr 应是一个合法的 URL 值
public URL(String protocol,String host, String file) throws MalformedURLException	创建一个拥有指定协议名称、主机名、文件名的 URL 对象
pubic URL(String protocol, String host, int port, String file) throws MaformedURLException	创建一个拥有指定协议名称、主机名、端口号、文件名的 URL 对象
public String getProtocol()	返回 URL 中的协议名称
public String getHost()	返回 URL 中的主机名
public int getPort()	返回 URL 中的端口号。如果 URL 中没有设定端口号，该函数返回-1
public String getFile()	返回 URL 中的文件名部分
public String getRef()	返回 URL 的引用
public String toString()	返回整个 URL 值

例 3-8　为 myweb 的下载页面创建一个 URL，然后检查它的属性。

```
import java.net.*;
public class Test
{
    public static void main(String args[]) throws MalformedURLException
    {
        URL hp = new URL("http://www.myweb.com/download");
        System.out.println("Protocol: " + hp.getProtocol());
        System.out.println("Port: " + hp.getPort());
        System.out.println("Host: " + hp.getHost());
        System.out.println("File: " + hp.getFile());
        System.out.println("Ext:" + hp.toExternalForm());
    }
}
```

为获 URL 的实际数据或内容信息，用它的 openConnection()方法由它创建一个 URLConnection 对象，常用形式为：

```
URLConnection openConnection()
```

它返回一个 URLConnection 对象并可能引发 IOException 异常。

URLConnection 是访问远程资源属性的一般用途的类。如果已建立了与远程服务器之间的连接，可以在传输它到本地之前用 URLConnection 来检察远程对象的属性。这些属性由 HTTP 规范定义，并且仅对用 HTTP 的 URL 对象有意义。

例 3-9　用 URL 对象的 openConnection()方法创建了一个 URLConnection 类，然后用它来检查文件的属性和内容。

```
import java.net.*;
import java.io.*;
import java.util.Date;
public class Test
{
    public static void main(String args[]) throws Exception
```

```
    {
        int c;
        URL hp = new URL("http://www.myweb.com");
        URLConnection hpCon = hp.openConnection();
        System.out.println("Date: " + new Date(hpCon.getDate()));
        System.out.println("Content-Type: " + hpCon.getContentType());
        System.out.println("Expires: " + hpCon.getExpiration());
        System.out.println("Last-Modified: "+new Date(hpCon.getLastModified()));
        int len = hpCon.getContentLength();
        System.out.println("Content-Length: " + len);
        if (len > 0)
        {
            System.out.println("Content: ");
            InputStream input = hpCon.getInputStream();
            int i = len;
            while (((c = input.read()) != -1) && (--i > 0))
                System.out.print((char) c);
            input.close();
        }
        else
            System.out.println("No Content Available");
    }
}
```

该程序建立了一个经过端口 80 通向 www.myweb.com 的 HTTP 连接，然后列出了标头值并检索内容。下面是输出的前几行：

```
Date: Fri Oct 06 22:11:12 CDT 2011
Content-Type: text/html
Expires: 0
Last-Modified: Tue Oct 28 11:16:27 CDT 2011
Content-Length: 529
Content:
<html>
<head>
<title>myweb </title>
</head>
```

3.3 基于 TCP/UDP 的编程

3.3.1 TCP 编程模型与实例

1. 编程模型

TCP/IP 套接字用于在主机和 Internet 之间建立可靠的、双向的、持续的、点对点的流式连接。一个套接字可以用来建立 Java 的输入/输出系统到其他的驻留在本地机或 Internet 上的任何机器的程序的连接。

需要注意的是，对于 applet 程序只建立回到下载它的主机的套接字连接。这个限制存在的原因是：穿过防火墙的 applet 程序有权使用任何机器是很危险的事情。

Java 中有两类 TCP 套接字，一类是服务器端的，另一类是客户端的。

Socket：客户端套接字类，为建立连向服务器套接字及启动协议交换而设计。

一个 Socket 对象的创建隐式建立了一个客户和服务器的连接，没有显式地说明建立连接细节的方法或构造方法。表 3.7 是用来生成客户套接字的两种构造方法。

表 3.7　客户端套接字类构造方法

方　法	含　义
Socket(String hostName, int port)	创建一个本地主机与给定名称的主机端口的套接字连接，可以引发一个 UnknownHostException 异常或 IOException 异常
Socket(InetAddress ipAddress, int port)	用一个预先存在的 InetAddress 对象和端口创建一个套接字，可以引发 IOException 异常

使用表 3.8 所示的方法，可以在任何时候检查套接字的地址和与之有关的端口信息。

表 3.8　地址与端口方法

方　法	含　义
InetAddress getInetAddress()	返回与 Socket 对象相关的 InetAddress
Int getPort()	返回与该 Socket 对象连接的远程端口
Int getLocalPort()	返回与该 Socket 连接的本地端口

一旦 Socket 对象被创建，同样可以检查它获得访问与之相连的输入和输出流的权力。如果套接字因为网络的连接中断而失效，表 3.9 所示的这些方法都能够引发一个 IOException 异常。

表 3.9　输入和输出流方法

方　法	含　义
InputStream getInputStream()	返回与调用套接字有关的 InputStream 类
OutputStream getOutputStream()	返回与调用套接字有关的 OutputStream 类
Void close()	关闭 InputStream 和 OutputStream

ServerSocket：服务器端套接字类，设计成在等待客户建立连接之前不做任何事的“监听器”。ServerSocket 类用来创建服务器，服务器监听本地或远程客户程序通过公共端口的连接。

ServerSocket 与通常的 Sockets 类完全不同。当创建一个 ServerSocket 类时，它在系统注册自己对客户连接感兴趣。ServerSocket 的构造方法反映了希望接收连接的端口号及希望排队等待上述端口的时间。队列长度告诉系统有多少与之连接的客户在系统拒绝连接之前可以挂起。队列的默认长度是 50。构造方法可以引发 IOException 异常。表 3.10 所示为服务器套接类构造方法。

表 3.10　服务器套接字类构造方法

方　法	含　义
ServerSocket(int port)	在指定端口创建队列长度为 50 的服务器套接字
ServerSocket(int port, int maxQueue)	在指定端口创建一个最大队列长度为 maxQueue 的服务器套接字
ServerSocket(int port, int maxQueue, InetAddress localAddress)	在指定端口创建一个最大队列长度为 maxQueue 的服务器套接字。在一个多地址主机上，localAddress 指定该套接字约束的 IP 地址

ServerSocket 有一个额外的 accept()方法，该方法是一个等待客户开始通信的模块化调用，然后以一个用来与客户通信的常规 Socket 返回。

基于 TCP 通信程序开发流程如图 3.3 所示。

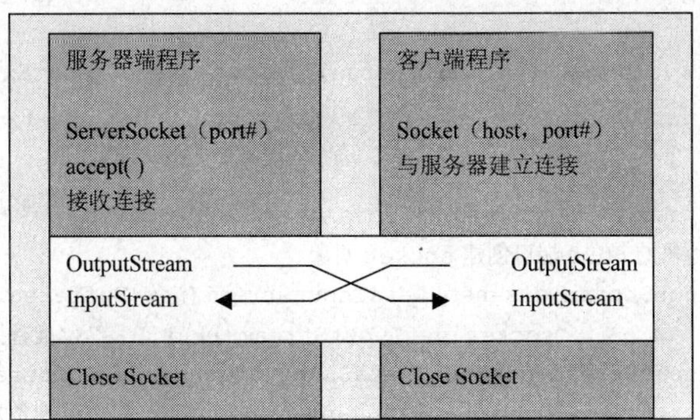

图 3.3　基于 TCP 通信程序开发

2. 编程实例

服务器端程序实现在端口 8080 上监听提供服务，从客户端读入数据后，再向客户端发送信息。同样，客户端发送数据后再接收服务器数据。该实例简单地实现两者的通信。

```
import java.net.*;
import java.io.*;
public class Server                                          //服务器端程序
{
   public static void main(String[] args)
   {
      try
      {
         ServerSocket ss=new ServerSocket(8080);
         //创建 ServerSocket 对象，端口为 8080
         while(true)
         {
            Socket socket=ss.accept();                       //接收客户端请求
            //监听客户端的请求，等待连接，连接完成返回 socket 套接字
            InputStream in=socket.getInputStream();          //定义输入流对象
            OutputStream out=socket.getOutputStream();  //定义输出流对象
            InputStreamReader reader=new InputStreamReader(in);
            BufferedReader bufReader=new BufferedReader(reader);
            String readLine=bufReader.readLine();            //读入一行
            System.out.println(readLine);
            PrintStream ps=new PrintStream(out);
            ps.print("Hello");                               //输出信息
            ps.flush();
            bufReader.close();
            ps.close();
            ss.close();
         }
      }
      catch(Exception e){
        System.out.println("Error");
      }
   }
}

import java.net.*;
```

```
import java.io.*;
public class Client                                                    //客户端程序
{
    public static void main(String[] args) throws UnknownHostException, IOException
    {
        try
        {
            Socket socket=new Socket("localhost",8080);
            //在本地机的 8080 端口创建 socket 对象
            DataInputStream dis=new DataInputStream(new BufferedInputStream(
                        socket.getInputStream()));      //准备接收服务器端信息
            DataOutputStream dos=new DataOutputStream(new BufferedOutputStream(
                        socket.getOutputStream()));     //准备发送数据对象
            dos.writeUTF("Hello Server!");              //发送数据
            dos.flush();
            System.out.println(dis.readUTF());
            dis.close();
            dos.close();
            socket.close();
        }
        catch(Exception e){
            System.out.println("Error");
        }
    }
}
```

3.3.2 UDP 编程模型与实例

1. 编程模型

对于当前的大多数网络需求，TCP/IP 型网络已经基本可以满足了。TCP/IP 型网络提供了有序的、可预测和可靠的信息包数据流。但是，这样做的代价也很大。TCP 包含很多在拥挤的网络中处理拥塞控制的复杂算法，以及信息丢失的悲观的预测。这导致了一个效率很差的传输数据方式。数据报是一种可选的替换方法。

数据报（Datagrams）是在机器间传递的信息包，一旦数据报被释放给它们预定的目标，不保证它们一定到达目的地，甚至不保证一定存在数据的接收者。同样，数据报被接收时，不保证它在传输过程不受损坏，不保证发送它的机器仍在等待响应。

Java 通过两个类实现 UDP 顶层的数据报。

DatagramPacket 类：创建一个用于发送的数据报，当接收数据 UDP 数据报时，可以使用 DatagramPacket 类读取数据报中的数据、发送者和其他消息。DatagramPacket 类的对象是数据容器。

DatagramSocket 类：封装了套接字的有关信息和操作发送端，是用来发送和接收 DatagramPackets 的机制。

构造方法：

```
public DatagramSocket() throws Excetpion
public DatagramSocket(int port) throws Exception
```

DatagramSocket()方法创建一个负责发数据报的 DatagramSocket 类对象，DatagramSocket(int port)创建一个指定了端口号的 DatagramSocket 类对象。

DatagramPackets 可以用 4 种构造方法中的一个创建。第 1 种构造方法指定了一个接收数据的缓冲区和信息包的容量大小，它通过 DatagramSocket 接收数据。第 2 种构造方法允许在存储数据的缓冲区中指定一个偏移量。第 3 种构造方法指定了一个用于 DatagramSocket 决定信息包将被送往何处的目标地址和端口。第 4 种构造方法从数据中指定的偏移量位置开始传输数据包。

下面是 4 种构造方法：

```
DatagramPacket(byte data[ ], int size)
DatagramPacket(byte data[ ], int offset, int size)
DatagramPacket(byte data[ ], int size, InetAddress ipAddress, int port)
DatagramPacket(byte data[ ], int offset, int size, InetAddress ipAddress, int port)
```

有几种方法可获取 DatagramPacket 内部状态，它们对信息包的目标地址和端口号，以及原始数据和数据长度有完全的使用权，表 3.11 所示简述了几种方法的含义。

表 3.11　DatagramPacket 方法

方　法	含　义
InetAddress getAddress()	返回目标文件 InetAddress，一般用于发送
Int getPort()	返回端口号
byte[ ] getData()	返回包含在数据包中的字节数组数据。多用于在接收数据之后从数据包来检索数据
Int getLength()	返回包含在将从 getData()方法返回的字节数组中的有效数据长度。通常它与整个字节数组长度不等

基于 UDP 通信程序的设计步骤如下。

数据报发送过程：

（1）创建 DatagramPacket 对象；

（2）在指定的或可用的本机端口创建 DatagramSocket 对象；

（3）调用该 DatagramSocket 对等的 send()方法，以 DatagramPacke 对象为参数发送数据报。

数据报接收过程：

（1）创建一个用于接收数据报的 DatagramPacket 对象，其中包含空白数据缓冲区和指定数据报分组的长度；

（2）在指定的或可用的本机端口创建 DatagramSocket 对象；

（3）调用 DatagramSocket 对象的 receive()方法，以 DatagramPacket 对象为参数接收数据报。

2. 编程实例

该实例实现通过 UDP 通信方式来进行发送和接收服务，即在本机端口 1010 上发生数据，在另一端接收数据。

```
import java.net.*;
import java.io.*;
public class UDPsender
{                                                              //发送数据
    public static void main(String[] args){
        byte sBuf[]=new byte[100];
        System.out.println("Input the message you will send: ");
        DataInputStream si=new DataInputStream(System.in);    //定义输入流对象
        try
        {
          int i;
```

```
            for(i=0;i<100;i++)
            {
               byte inByte= si.readByte();
               if ((char)inByte=='#')              //判断输入是否为结束符#
                  break;
               sBuf[i]=inByte;
            }
            DatagramSocket sk=new DatagramSocket();  //生成 DatagramSocket 对象
            DatagramPacket packet=new DatagramPacket(
                        sBuf,i,InetAddress.getByName("localhost"),1010);
                        //生成数据包装类对象
            sk.send(packet);                                //发送数据
            sk.close();
         }
         catch(Exception e){
           System.out.println("Error");
         }
      }
   }
   public class UDPReceive
   {                                                         //接收数据
       public static void main(String[] args)
       {
           byte rBuf[]=new byte[100];
           DatagramPacket packet=new DatagramPacket(rBuf,rBuf.length);
          //生成数据包装类
           try
           {
              DatagramSocket rs=new DatagramSocket(1010); //生成 DatagramSocket 对象
              rs.receive(packet);                            //接收数据
              System.out.println(new String(packet.getData()); //输出数据
              rs.close();
           }
           catch(Exception e)
           {
               System.out.println("Error");
           }
       }
   }
```

思考与习题

1. 填空题：TCP 是________协议，它主要负责________________。

2. 填空题：UDP 是________________协议，它和 TCP 一样都是网络传输层上的协议。

3. 填空题：常用的 TCP/IP 的 3 种套接字类型，即________________、数据报套接字和________________。

4．填空题：Socket 套接字是为________________________________而设计。

5．填空题：在 UDP 编程模型中，Java 通过两个类实现 UDP 顶层的数据报：________________和________________。

6．简答题：简述进程与线程的区别。

7．简答题：简述客户/服务器模式，以及其基于套接字的通信方式。

8．编程题：用线程模拟队列报数，要求第 1 个队列报完数后第 2 个队列再报数。

9．编程题：将客户端录入的 10 个数传送到服务器端，计算平均值结果再传送给客户端。

10．编程题：将某个网站的首页内容输出到控制台。

第4章 Java图形用户界面

图形用户界面或图形用户接口（Graphical User Interface，GUI）是指采用图形方式显示的计算机操作环境用户接口。与早期计算机使用的命令行界面相比，图形界面对于用户来说更为简便易用。

本章首先概述了图形用户界面及其组件的基本知识，接着介绍Java中常用的Swing组件和布局管理器，以及介绍与Java图形用户接口相互作用时产生的事件（如在界面中通过按钮、鼠标等），最后给出了一个公司通信录系统的界面编程实例。

4.1 概述

4.1.1 图形用户界面

图形用户界面为用户与程序友好地交互提供了一种机制，由标签、文本、单选按钮、复选框、列表框、窗口等组成，用户通过鼠标、键盘或其他方式与这些组件交互。

Java提供了一系列的用户界面组件，这些组件主要包含在两个包中，一个是抽象窗口工具包（AWT），包名为java.awt；另一个是Swing包，包名为java.swing。在编程应用中，由于Java是构建面向网络的应用，而该包不具有实际的编程价值。对于初学Java技术的人员来说，本章具有引导学习的作用。

在Java中，图形用户界面由组件、布局管理和事件委托处理3部分组成。组件是响应用户操作的可视图形控件，它们置于容器组件中形成一个具有包含关系的层次结构，由布局管理器安排组件在容器中的位置。能够产生事件的组件称事件源，不同的组件产生不同类型的事件，同一组件可能产生多种事件。组件提供添加监视事件发生的监视器方法，对感兴趣的事件注册监听。将一个包含相关信息的事件对象传递给事件处理程序，即为提供相应的方法对事件进行业务处理。

创建一个图形用户界面的方法如下。

（1）创建一个顶级的容器组件，一般是继承框架（Frame）的应用程序主窗口，它包含边框、标题栏、控制菜单和内容面板（Contentpane）的容器。

（2）按指定的布局管理器向容器中添加组件，给事件源组件添加事件监视器，处理注册事件。

（3）设计事件处理程序，重新处理方法。

（4）显示图形用户界面。

4.1.2　组件

组件是抽象的概念，通俗地说是一些符合某种规范的类或代码组合在一起就构成了组件，对外提供多个接口或方法供调用，从而完成某些特定的功能。

Java 中所有的 UI 组件都是以 Component 类为根。抽象窗口工具包（Abstract Window Toolkit，AWT）API 为 Java 程序提供的建立图形用户界面（Graphics User Interface，GUI）工具集，AWT 可用于 Java 的 applet 和 application 中。它支持图形用户界面编程的功能，包括组件生成、事件处理、图形和图像工具、布局管理等。AWT 组件采用底层的技术实现，具有平台相关性，是重量级组件。

Swing 是 AWT 的扩展，它提供了更强大和更灵活的组件集合。除按钮、复选框、标签等组件外，Swing 还包括许多新的组件，如选项板、滚动窗口、树和表格。许多组件，如按钮，在 Swing 中都增加了新功能。大部分的 Swing 组件都是以 JComponent 为根，这些类基本上都在 javax.swing 包中。

与 AWT 组件不同，Swing 组件的实现不包括任何与平台相关的代码。Swing 组件是纯 Java 代码，与平台无关，保证代码的可移植性。在 Swing 组件中，除了 JFrame、JWindow、JDialog 和 Japplet 外，都属于轻量级组件。上述这 4 个组件属于重量级组件，但它们与 AWT 的对应组件存在着区别。这 4 个组件属于窗体框架，可以在其中放置其他的组件，但与 AWT 不同，被放置的组件不是直接放在这 4 个框架上，而是放在框架中的内容窗格中。

在 javax.swing 包中，定义了两种类型的组件：顶层容器（Jframe、Japplet、JDialog 和 JWindow）和轻量级组件。Swing 组件都是 AWT 的 Container 类的直接子类和间接子类。

```
java.awt.Component
        -java.awt.Container
                -java.awt.Window
                        -java.awt.Frame-javax.swing.JFrame
                        -javax.Dialog-javax.swing.JDialog
                        -javax.swing.JWindow
                -java.awt.Applet-javax.swing.JApplet
                -javax.swing.Box
                -javax.swing.Jcomponet
```

例 4-1　引入 awt 和 swing 组件。

```
import java.awt.*;
import javax.swing.*;
class Test extends JFrame
{
    public void FirstFrame(String title)
    {
        super(title);
        this.setSize(400,300);
        this.setVisible(true);
        this.setDefaultCloseOperation(JFrame.EXIT_ON_CLOSE);
    }
}
public class MainApp
{
    public static void main(String[] args)
    {
        new Test ("GUI 演示");
    }
}
```

4.2 Swing 组件

AWT（AbstractWindowToolkit）为抽象窗口工具包，是 Java 提供的用来建立和设置 Java 的图形用户界面的基本工具。AWT 是 Swing 的基础，Swing 比 AWT 提供了更多的组件和外观，它能够满足 AWT 不能满足的图形化用户界面发展的需要。

Swing 组件的实现不包括任何与平台相关的代码，由于它是纯 Java 代码，因此它与平台是无关的。下面将通过文本、按钮、列表等介绍 Swing 组件。

4.2.1 文本组件

下面具体介绍 3 种文本组件，即 JLable、JtextField 和 JTextArea。

1. JLable

构造方法：

```
JLabel()
JLabel(Icon image)
JLabel(String text)
JLabel(Icon image, int horizontalAlignment)
JLabel(String text, int horizontalAlignment)
JLabel(String text, Icon icon, int horizontalAlignment)
```

主要方法：

```
public void setText(String text)
public String getText(String text)
public void setIcon(Icon icon)
public Icon getIcon()
```

2. JTextField

JTextField 与 JTextArea 组件都比较复杂。这个组件是文本框，只能输入单行的文本。当文本区域的大小超出了可显示的范围时，默认不会自动出现滚动条。

构造方法：

```
JTextField()
JTextField(Document doc, String text, int columns)
JTextField(int columns)
JTextField(String text)
JTextField(String text, int columns)
```

主要方法：

```
public void addActionListener(ActionListener listner)
public void setText(String text)
public String getText()
```

3. JTextArea

JTextArea 是文本区域，可以输入多行文本。当文本的区域超出可显示的范围时，在默认情况下，不会自动出现滚动条。

构造方法：

```
JTextArea()
JTextArea(Document doc)
JTextArea(Document doc, String text, int rows, int columns)
JTextArea(int rows, int columns)
JTextArea(String text)
JTextArea(String text, int rows, int columns)
```

主要方法：

```
public void append(String text)
public void setText(String text)
public String getText()
```

4.2.2 按钮组件

按钮组件是以 AbtractButton 为父类。当在这些组件上单击鼠标或在这些组件获得焦点时按 Enter 键，都会产生 ActionEvent 事件对象。这些事件对象能够被 ActionListener 监听器对象接收到，前提条件就是该监听器对象必须向这些按钮组件对象注册。

1. JButton()

它的构造方法如下：

```
JButton()
JButton(Action a)
JButton(Icon icon)
JButton(String text)
JButton(String text,Icon icon)
```

2. JCheckBox 和 JRadioButton

这两个组件对象都有一个是否被选中的状态。通常利用状态变化时，它们会产生 ItemEvent 事件的特点来帮助达到目的。而 ItemEvent 事件对象可以被 ItemListener 监听器对象接收到，前提条件就是将监听器对象向事件源对象注册。

```
public void addItemListener(ItemListener listener)
```

对于 JCheckBox 对象来说，一组这种类型的对象可以全部被选中，也可以全部不选中，或者部分选中；而对于一组 JRadioButton 对象来说，有且仅有一个对象被选中。在默认的情况下，每一个 JRadionButton 对象组成一组，如果想让几个 JRadioButton 对象组成为一组，那么必须使用 ButtonGroup 对象。

4.2.3 列表组件

1. JComboBox

组合框由下拉列表与文本框组成，可以通过 setEnable 方法来设置该组合框对象能否被编辑；通过 addItem 方法向该组合框对象增加项目；通过 getSelectedIndex 方法获得选择到的项目的索引号；通过 getSelectedItem 方法获得选择到的项目；也可以在获得项目索引的前提下，通过 getItemAt 方法获得选择到的项目。

在编程时，经常会用到该组件产生的 ItemEvent 事件，这个事件是在项目被选择到或已被选择到的项目变化不被选择时发生。这个事件可以被 ItemListener 监听器对象接收到。所以，下面相应的注册方法在编程时经常会被用到：

```
public void addItemListener(ItemListener listener)
```

2. JList

JList 列表框组件比较复杂，它可以多选，也可以单选，还可以通过 setSelectionMode 方法来实现，该方法的参数有效值有如下 3 个。

ListSelectionModel.SINGLE_SELECTION：只能一次选择一个项目。

ListSelectionModel.SINGLE_INTERVAL_SELECTION：一次可以只选择一个项目，也可以一次同时选择连续的多个项目（通过按 Shift 键），如果通过同时按下 Ctrl 键，那么必须是连续的项目才能被选中。

ListSelectionModel.MULTIPLE_INTERVAL_SELECTION：这是默认的选择模式，允许一次选择不连续的多个项目。当选择到项目时，该组件对象会产生 ListSelectionEvent 事件对象，这样的事件对象可以被 ListSelectionListener 监听器对象接收到。但前提条件是必须事先注册。列表框对象的注册 ListSelectionListener 监听器对象的方法如下：

```
public void addListSelectionListener(ListSelectionListener listener)
```

4.3 布局管理器

在 Swing 组件的 GUI 应用程序中，作为框架用的有 4 个组件，分别是 JFrame、JApplet、Jwindow 和 Jdialog。这 4 个组件中都包含了可以放置其他组件的容器 Container，所以，这 4 个组件都有一个返回该容器的方法 getContentPane()。对于 Swing 组件来说，其中有一些是作为可以放置其他组件的容器，如 JPanel。对于所有的容器来说，都有一个默认的布局管理器对象来管理放置在其中的组件，同时也可以修改布局管理器对象，这是通过 setLayout（LayoutManager mgr）方法来实现的，其中的参数就是布局管理器的最高类（抽象类）。

当一个组件加入一个容器时，组件可以向与该容器相关的布局管理器对象提供自己希望的尺寸和排列方式。但是这些组件的实际尺寸和位置最终还是由布局管理器对象来确定的。

对于 4 个框架组件来说，其中包含容器的默认布局管理器是 BorderLayout；对于 JPanel 容器来说，其默认的布局管理器是 FlowLayout。

4.3.1 顺序布局

布局管理器采用从左向右、从上向下的顺序来排列其中的组件，如果一行已满，则开始下一行。在默认的情况下，采用居中对齐的方式来排列组件，组件间的间隔为 5 个像素。这个布局管理器适用于组件个数较少的情况，如图 4.1 所示。

图 4.1 FlowLayout 顺序布局

可以在使用 FlowLayout 布局管理器构造对象时，使用常量参数来指定组件对齐方式（靠左、靠右、居中）。FlowLayout 共定义了 5 个常量用来指明组件对齐方式。

FlowLayout.LEADING：与开始一边对齐。

FlowLayout.TRAILING：与结束一边对齐。

FlowLayout.CENTER：居中对齐。

FlowLayout.LEFT：左对齐。

FlowLayout.RIGHT：右对齐。

可以使用这些常量来指定组件的对齐方式，如

```
setLayout(new FlowLayout(FlowLayout.LEFT,10,20));
```

4.3.2　边框布局

BorderLayout 是一个区域布局，它将整个容器分为东、西、南、北中 5 个区域，每个区域加入一个组件，如果在同一个区域加入两个或两个以上的组件时，最后加入的组件将覆盖前面加入的组件。在加入组件时，需要指明将要加入的区域。

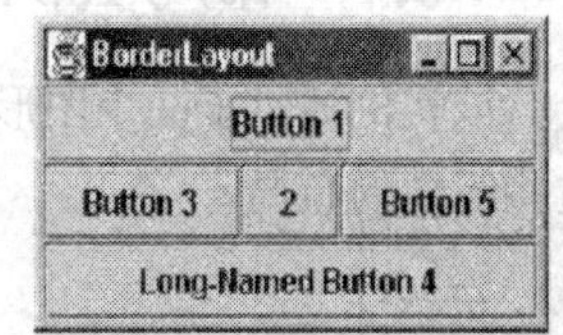

图 4.2　BorderLayout 边框布局

这个布局管理器最多使容器能容纳 5 个组件，如图 4.2 所示。

一个 BorderLayout 有 5 个区域，这 5 个区域由 BorderLayout 中的常量来指定：

```
PAGE_START;
PAGE_END;
LINE_START;
LINE_END;
CENTER;
```

如果窗口扩大，中间区域会尽可能多地占据空间，而其他 4 个区域只根据需要扩展以填充可用的空间。在很多情况下，容器只会使用到 BorderLayout 对象中的一个或两个区域，如 CENTER 或 CENTER 和 PAGE_END。

默认情况下，内容面板使用到 BorderLayout 布局。

4.3.3　网格布局

GridLayout 是一个网格布局，它将整个容器分为 n 行 m 列共 $n \times m$ 个大小相同的小格。在默认情况下，小格之间的间隔为 0。每一个小格可以放置一个组件。当向容器加入组件时，按从左向右、从上向下的顺序依次放置，如图 4.3 所示。

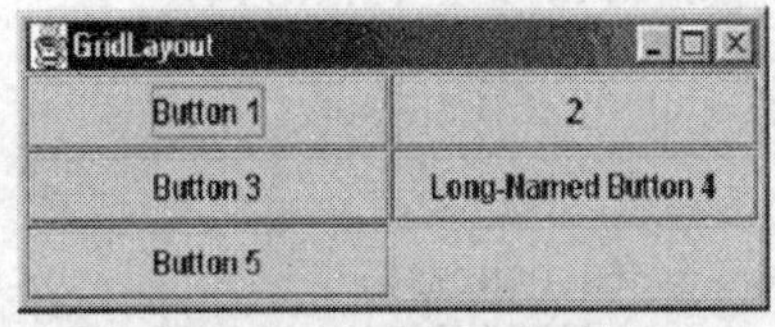

图 4.3　GridLayout 网格布局

可以通过构造方法来设置小格的数量和小格之间的空隙。例如，new GridLayout(5,10,2,4); 创建一个 5 行 10 列的网格，每个小格之间的水平间隔为 2 个像素，垂直间隔为 4 个像素。当然也可以采用默认的间隔，如：new GridLayout(5,10); 创建一个 5 行 10 列的网格，每个小格之间的间隔为 0。

GridLayout 的布局方式将容器按用户的设置平均划分成若干个网格，其中的每一格称为一个“单元格”。在通过构造方法 GridLayout(int rows,int cols)　创建网格布局管理器对象时，参数 rows 用来设置网格的行数，参数 cols 用来设置网格的列数。在设置时分为以下 4 种情况。

（1）只设置了网格的行数，即 rows 大于 0，cols 等于 0。在这种情况下，容器将先按行排列组件，当组件数大于 rows 时，再增加一列。依此类推。

（2）只设置了网格的列数，即 cols 大于 0，rows 等于 0。在这种情况下，容器将先按列排列组件，当组件数大于 cols 时，再增加一行。依此类推。

（3）同时设置了行数和列数，即 rows 大于 0，cols 大于 0。在这种情况下，容器将先按行排列组件，当组件数大于 rows 时，再增加一列。依此类推。

（4）同时设置了行数和列数，但组件数大于网格数（rows × cols）。在这种情况下，容器将先按行排列组件，当组件数大于 rows 时，再增加一列，依此类推。

GridLayout 布局管理器默认组件的水平间距和垂直间距均为 0 像素，可以通过 GridLayout 类的方法 setHgap（int hgap）和 setVgap（int vgap）设置组件的水平间距和垂直间距。

4.3.4 布局实例

该实例演示了一个图形界面的实现，有窗口、标签、按钮、输入文本框等组件，同时规定了这些组件的布局。

```
import java.awt.*;
import java.awt.event.*;
import javax.swing.*;                                             //导入 swing 包
class MessagePanel extends JPanel                                 //定义面板
{
    private JLabel label;
    public MessagePanel(String msg)
    {
        setLayout(new GridLayout(1,1));                           //设置网格布局
        label=new JLabel(msg);
        label.setHorizontalAlignment(SwingConstants.CENTER);      //设置对齐方式
        label.setFont(new Font("宋体",Font.BOLD,20));              //设置字体
        add(label);
    }
}
class ArithPanel extends JPanel
{
    private JLabel[] labels;
    private JTextField result;
    public ArithPanel()
    {
        setLayout(new GridLayout(1,5,1,1));                       //设置网格布局
        labels=new JLabel[4];
        for(int i=0;i<labels.length;i++)
        {
            labels[i]=new JLabel();
            labels[i].setOpaque(true);
            labels[i].setHorizontalAlignment(SwingConstants.CENTER);
            labels[i].setBackground(Color.white);         //设置背景颜色
            add(labels[i]);
        }
        labels[0].setText("12");
        labels[1].setText("+");
        labels[2].setText("88");
        labels[3].setText("=");
        result=new JTextField();                                  //生成文本输入框
        add(result);                                              //加入面板
```

```
    }
}
class NorthPanel extends JPanel
{
        private MessagePanel msg;
        private ArithPanel arith;
    public NorthPanel()
    {
            msg=new MessagePanel("好好学习！");
            arith=new ArithPanel();
            setLayout(new GridLayout(2,1));
            add(msg);
            add(arith);
    }
}
class ButtonsPanel extends JPanel
{
        private JButton start;
        private JButton show;
        private JButton exit;
        public ButtonsPanel()
        {
             start=new JButton("开始");          //定义按钮
             show=new JButton("显示");
             exit=new JButton("退出");
             show.setEnabled(false);            //设置可用性
             exit.addActionListener(            //按钮增加事件监听器
                  new ActionListener()          //事件监听器对象
                  {
                      public void actionPerformed(ActionEvent e)
                      {                         //事件执行体
                          System.exit(0);       //退出程序
                      }
                  });
            add(start);
            add(show);
            add(exit);
        }
}
public class ArithmeticFrame extends JFrame
{                                               //定义窗口容器类
    private JTextArea area;
        public ArithmeticFrame()                //构造方法
        {
            NorthPanel north=new NorthPanel();
            area=new JTextArea();
            ButtonsPanel south=new ButtonsPanel();
            Container c=this.getContentPane(); //获取内容面板
            c.add(north,BorderLayout.NORTH);   //将面板加入容器的一个位置
            c.add(new JScrollPane(area),BorderLayout.CENTER);
            c.add(south,BorderLayout.SOUTH);
            this.setSize(300,400);             //设置窗口大小
            this.setVisible(true);             //设置可见
```

```
            this.setDefaultCloseOperation(JFrame.EXIT_ON_CLOSE);
    }
        public static void main(String[] args)
    {
        new ArithmeticFrame();
    }
}
```

应用程序的运行结果如图 4.4 所示。

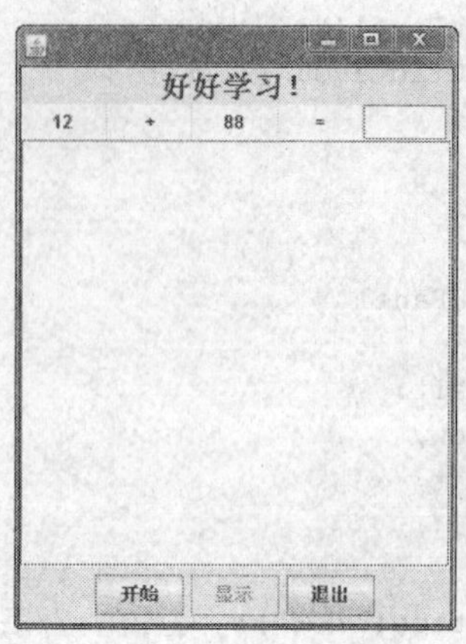

图 4.4　运行结果

4.4　事件

在授权事件模型中，一个事件是一个描述了事件源的状态改变的对象，它可以作为一个人与图形用户接口相互作用的结果被产生。本节主要介绍事件的处理机制和典型的鼠标键盘事件、事件源监听器的基本概念和事件处理的实现等。

4.4.1　事件处理机制

Java 中处理事件的方法是基于授权事件模型的，这种模型定义了标准一致的机制去产生和处理事件。它的概念十分简单：一个源产生一个事件，并把它送到一个或多个监听器那里。在这种方案中，监听器简单地等待，直到它收到一个事件。一旦事件被接收，监听器将处理这些事件，然后返回。这种设计的优点是那些处理事件的应用程序可以明确地和那些用来产生那些事件的用户接口程序分开。一个用户接口元素可以授权一段特定的代码处理一个事件。

在授权事件模型中，监听器为了接收一个事件通知必须注册。这样有一个重要的好处，就是通知只被发送给那些想接收它们的监听器那里。

在授权事件模型中，一个事件是一个描述了事件源的状态改变的对象。它可以作为一个人与图形用户接口相互作用的结果被产生，如一些产生事件的活动可以是通过按一个按钮、用键盘输入一个字符、选择列表框中的一项、单击一下鼠标等。同时，事件也可能不是由于用户接口的交互而直接发生的，如一个事件可能由于定时器到期，一个计数器超过了一个值，一个软件或硬件错误发生，或者一个操作被完成而产生。用户还可以自由地定义一些适用于自己的应用程序的事件。

事件源是一个产生事件的对象，当这个对象内部的状态以某种方式改变时，事件就会产生。事件源可能产生不止一种事件。一个事件源必须注册监听器以便监听器可以接收关于一个特定事件的通知。每一种事件有它自己的注册方法，通用的形式为：

```
public void addTypeListener(TypeListenerel)
```

在这里 type 是事件的名称，而 el 是一个事件监听器的引用。例如，注册一个键盘事件监听器的方法被叫作 addKeyListener()，注册一个鼠标活动监听器的方法被叫作 addMouseMotionListener()，当一个事件发生时，所有被注册的监听器都被通知并收到一个事件对象的复制。在所有情况下，事件通知只被送给那些注册接收它们的监听器。

一些事件源可能只允许注册一个监听器。一个事件源必须也提供一个允许监听器注销一个特定事件的方法，该方法的通用形式如下：

```
public void removeTypeListener(TypeListener el)
```

这里 type 是事件的名字，而 el 是一个事件监听器的引用。例如，为了注销一个键盘监听器，将调用 removeKeyListener()方法。

这些增加或删除监听器的方法由产生事件的事件源提供。例如，component 类提供了那些增加或删除键盘和鼠标事件监听器的方法。

事件监听器是一个在事件发生时被通知的对象。它有两个要求，首先，为了可以接收到特殊类型事件的通知，它必须在事件源中已经被注册；其次，它必须实现接收和处理通知的方法。

用于接收和处理事件的方法在 java.awt.event 中被定义为一系列的接口。例如，MouseMotionListener 接口定义了两个在鼠标被拖动时接收通知的方法。如果实现这个接口，任何对象都可以接收并处理这些事件的一部分。

4.4.2 鼠标和键盘事件

Java 事件处理机制的核心是代表这些事件的类。事件类提供一个统一而又易用的封装事件的方法。表 4.1 列举了这些事件类中最重要的一些事件类，并对它们的产生条件进行了简要描述。

表 4.1 主要事件类

事 件 类	描 述
ActionEvent	通常在按下一个按钮，双击一个列表项或者选中一个菜单项时发生
AdjustmentEvent	当操作一个滚动条时发生
ComponentEvent	当一个组件隐藏、移动、改变大小或成为可见时发生
ContainerEvent	当一个组件从容器中加入或删除时发生
FocusEvent	当一个组件获得或失去键盘焦点时发生
InputEvent	所有组件的输入事件的抽象超类
ItemEvent	当一个复选框或列表项被点击时发生；当一个选择框或一个可选择菜单的项被选择或取消时发生
KeyEvent	当输入从键盘获得时发生
MouseEvent	当鼠标被拖动、移动、单击、按下、释放时发生，或者在鼠标进入或退出一个组件时发生
TextEvent	当文本区和文本域的文本改变时发生
WindowEvent	当一个窗口激活、关闭、失效、恢复、最小化、打开或退出时发生

1. MouseEvent 类

鼠标事件可通过接口 MouseMotionListener 和接口 MouseListener，关于这两个接口将在下节做具体介绍。响应鼠标事件有 7 种类型，在 MouseEvent 类中定义了表 4.2 所示的整型常量。

表 4.2　鼠标事件整型常量

整 型 常 量	描　述
MOUSE_CLICKED	用户单击鼠标
MOUSE_DRAGGED	用户拖动鼠标
MOUSE_ENTERED	鼠标进入一个组件内
MOUSE_EXITED	鼠标离开一个组件
MOUSE_MOVED	鼠标移动
MOUSE_PRESSED	鼠标被按下
MOUSE_RELEASED	鼠标被释放

MouseEvent 类是 InputEvent 类的子类，它的构造方法如下：

```
MouseEvent(Component src, int type, long when, int modifiers, int x, int y, int clicks,
boolean triggersPopup);
```

这里 src 是一个产生事件的组件的引用，type 指定了事件的类型，鼠标事件发生时的系统时间在 when 中被传递，参数 modifiers 决定了在鼠标事件发生时哪一个修改键被按下，鼠标的坐标在 x、y 中传递，单击的次数在 clicks 中传递，triggersPopup 标志决定了是否由这个事件引发在平台上弹出一个弹出式菜单。

在这个类中用得最多的方法是 getX()和 getY()。它们返回事件发生时鼠标所在坐标点的 X 和 Y，其形式如下：

```
int getX();
int getY();
```

相应地，也可以用 getPoint()方法去获得鼠标的坐标，形式如下：

```
Point getPoint();
```

它返回了一个 Point 对象，在这个对象中以整数成员变量的形式包含了 x 和 y 坐标。

translatePoint()方法可以改变事件发生的位置，它的形式如下：

```
void translatePoint(int x, int y);
```

这里参数 x 和 y 被加到了该事件的坐标中。

getClickCount()方法可以获得这个事件中鼠标的单击次数，如下所示：

```
int getClickCount();
```

isPopupTrigger()方法可以测试是否这个事件将引起一个弹出式菜单在平台中弹出，如下所示：

```
boolean isPopupTrigger();
```

2. KeyEvent 类

键盘事件是当键盘输入时产生的事件。键盘事件有 3 种，它们分别用整型常量 KEY_PRESSED、KEY_RELEASED 和 KEY_TYPED 来表示。前两个事件在任何键被按下或释放时发生，而最后一个事件只在产生一个字符时发生。请记住，不是所有被按下的键都产生字符。例如，按 Shift 键就不能产生一个字符。

还有许多整型常量在 KeyEvent 类中被定义。例如，从 VK_0 到 VK_9 和从 VK_A 到 VK_Z 定义了与这些数字和字符等价的 ASCII。还有一些其他的常量如 VK_ENTER、VK_ESCAPE、VK_CANCEL、VK_UP、VK_DOWN、VK_LEFT、VK_RIGHT、VK_PAGE_DOWN、VK_PAGE_UP、VK_SHIFT、VK_ALT、VK_CONTROL 等。VK 常量指定了虚拟键值（virtual key codes），并且与任何 Ctrl、Shift 或 Alt 修改键不相关。

KeyEvent 类是 InputEvent 类的子类，它有两种构造方法：

```
KeyEvent(Component src, int type, long when, int modifiers, int code);
KeyEvent(Component src, int type, long when, int modifiers, int code, char ch);
```

这里 src 是一个产生事件的组件的引用，Type 指定了事件的类型。当这个键被按下时，系统时间在 when 里被传递。参数 modifiers 决定了在键盘事件发生时哪一个修改符被按下。像 VK_UP 和 VK_A 这样的虚拟键值在 code 中传递。如果与这些虚拟键值相对应的字符存在，则在 ch 中被传递，否则 ch 中是 CHAR_UNDEFINED。对于 KEY_TYPED 事件，code 将是 VK_UNDEFINED。

KeyEvent 类定义了一些方法，其中用得最多的是用来返回一个被输入的字符的方法和返回键值的方法 getKeyCode()。它们的通常形式如下：

```
char getKeyChar();
int getKeyCode();
```

如果没有合法的字符可以返回，getKeyChar()方法将返回 CHAR_UNDEFINED。同样，在一个 KEY_TYPED 事件发生时，getKeyCode()方法返回的是 VK_UNDEFINED。

4.4.3 事件源和监听器

表 4.3 中列举了一些可以产生我们在前面所描述的事件的用户接口组件。当然，除了这些图形用户接口元素之外，其他组件，如一个 applet 程序，也可以产生事件。例如，可以在一个 applet 程序中获得键盘和鼠标事件。

表 4.3 事件源举例

事 件 源	描 述
Button	在按钮被按下时产生动作事件
Checkbox	在复选框被选中或取消时产生项目事件
Choice	在选择改变时产生项目事件
List	在一项被双击时，产生动作事件，被选择或取消时产生项目事件
Menu item	菜单项被选中时产生动作事件，当可复选菜单项被选中或取消时产生项目事件
Scrollbar	在滚动条被拖动时产生调整事件
Text components	当用户输入字符时产生文本事件
Window	窗口被激活、关闭、失效、恢复、最小化、打开或退出时产生窗口事件

正如我们前面所解释的，在授权事件模型中有两部分：事件源和监听器。事件源是通过实现一些在 java.awt.event 包中被定义的接口而生成的。当一个事件产生的时候，事件源调用被监听器定义的相应的方法，并提供一个事件对象作为参数。在表 4.4 中列出了通常用到的监听器接口，同时还简要地说明了它们所定义的方法。接下来将解释每一个接口包含的一些特殊方法。

表 4.4　通常使用的事件监听器接口

接　口	描　述
ActionListener	定义了 1 种接收动作事件的方法
AdjustmentListener	定义了 1 种接收调整事件的方法
ComponentListener	定义了 4 种方法来识别何时隐藏、移动、改变大小、显示组件
ContainerListener	定义了 2 种方法来识别何时从容器中加入或除去组件
FocusListener	定义了 2 种方法来识别何时组件获得或失去焦点
ItemListener	定义了 1 种方法来识别何时项目状态改变
KeyListener	定义了 3 种方法来识别何时键按下、释放和输入字符事件
MouseListener	定义了 5 种方法来识别何时鼠标单击、进入组件、离开组件、按下和释放事件
MouseMotionListener	定义了 2 种方法来识别何时鼠标拖动和移动
TextListener	定义了 1 种方法来识别何时文本值改变
WindowListener	定义了 7 种方法来识别何时窗口激活、关闭、失效、最小化、还原、打开和退出

1. ActionListener 接口

在这个接口中定义了 actionPerformed()方法，当一个动作事件发生时，它将被调用。一般形式如下：

```
void actionPerformed(ActionEvent ae);
```

2. AdjustmentListener 接口

在这个接口中定义了 adjustmentValueChanged()方法，当一个调整事件发生时，它将被调用。一般形式如下：

```
void adjustmentValueChanged(AdjustmentEvent ae);
```

3. ComponentListener 接口

在这个接口中定义了 4 种方法，当一个组件被改变大小、移动、显示或隐藏时，它们将被调用。一般形式如下：

```
void componentResized(ComponentEvent ce);
void componentMoved(ComponentEvent ce);
void componentShown(ComponentEvent ce);
void componentHidden(ComponentEvent ce);
```

注意：AWT 处理改变大小和移动事件。componentResized()和 componentMoved()方法只用来提供通知。

4. ContainerListener 接口

在这个接口中定义了两个方法，当一个组件被加入到一个容器中时，componentAdded()方法将被调用。当一个组件从一个容器中删除时，componentRemoved()方法将被调用。这两种方法的一般形式如下：

```
void componentAdded(ContainerEvent ce);
void componentRemoved(ContainerEvent ce);
```

5. FocusListener 接口

在这个接口中定义了两个方法，当一个组件获得键盘焦点时，focusGained()方法将被调用。当一个组件失去键盘焦点时，focusLost()方法将被调用。这两个方法的一般形式如下：

```
void focusGained(FocusEvent fe);
void focusLost(FocusEvent fe);
```

6. ItemListener 接口

在这个接口中定义了 itemStateChanged()方法，当一个项的状态发生变化时它将被调用。这个方法的原型如下：

```
void itemStateChanged(ItemEvent ie);
```

7. KeyListener 接口

在这个接口中定义了 3 种方法。当一个键被按下和释放时，相应地 keyPressed()方法和 keyReleased()方法将被调用。当一个字符已经被输入时，keyTyped()方法将被调用。例如，如果一个用户按下和释放 A 键，通常有 3 个事件顺序产生：键被按下、输入和释放。

如果一个用户按下和释放 Home 键时，通常有两个事件顺序产生：键被按下和释放。这些方法的一般形式如下：

```
void keyPressed(KeyEvent ke);
void keyReleased(KeyEvent ke);
void keyTyped(KeyEvent ke);
```

8. MouseListener 接口

在这个接口中定义了 5 种方法。当鼠标在同一点被按下和释放时，mouseClicked()方法将被调用；当鼠标进入一个组件时，mouseEntered()方法将被调用；当鼠标离开组件时，mouseExited()方法将被调用；当鼠标被按下和释放时，相应的，mousePressed()方法和 mouseReleased()方法将被调用。

这些方法的一般形式如下：

```
void mouseClicked(MouseEvent me);
void mouseEntered(MouseEvent me);
void mouseExited(MouseEvent me);
void mousePressed(MouseEvent me);
void mouseReleased(MouseEvent me);
```

9. MouseMotionListener 接口

在这个接口中定义了两个方法，当鼠标被拖动时，mouseDragged()方法将被调用多次。当鼠标被移动时，mouseMoved()方法将被调用多次。这些方法的一般形式如下：

```
void mouseDragged(MouseEvent me);
void mouseMoved(MouseEvent me);
```

10. TextListener 接口

在这个接口中定义了 textChanged()方法，当文本区或文本域发生变化时，它将被调用。这个方法的一般形式如下：

```
void textChanged(TextEvent te);
```

11. WindowListener 接口

在这个接口中定义了 7 个方法。当一个窗口被激活或禁止时，windowActivated()方法或 windowDeactivated()方法将相应地被调用；如果一个窗口被最小化，windowIconified()方法将被调用；当一个窗口被恢复时，windowDeIconified()方法将被调用；当一个窗口被打开或关闭时，windowOpened()方法或 windowClosed()方法将相应地被调用；当一个窗口正在被关闭时，windowClosing()方法将被调用。

这些方法的一般形式如下：

```
void windowActivated(WindowEvent we);
void windowClosed(WindowEvent we);
void windowClosing(WindowEvent we);
void windowDeactivated(WindowEvent we);
void windowDeiconified(WindowEvent we);
void windowIconified(WindowEvent we);
void windowOpened(WindowEvent we);
```

4.4.4 事件处理实现

利用授权事件模型来编写简单的 GUI 应用程序，只需要如下两步。

（1）在监听器中实现相应的监听器接口，以便接收相应的事件。

（2）实现注册或注销（如果必要）监听器的代码，以便可以得到事件的通知。

请记住，一个事件源可能产生多种类型的事件，每一个事件都必须分别注册。当然，一个对象可以注册接收多种事件，但是它必须实现相应的所有事件监听器的接口。

为了明确授权事件模型实际上是如何工作的，下面将分析一个例子，在这个例子中处理了两个最常用的事件产生器：鼠标和键盘。

如果要处理鼠标事件，就必须实现 MouseListener 接口和 MouseMotionListener 接口。下面的程序在窗体上显示了鼠标的当前坐标。每当鼠标按钮被按下，在鼠标指针所在的位置将显示 “按下”。而当鼠标按钮释放时，将显示“释放”。如果鼠标按钮被单击，将显示“单击”。

当鼠标进入或退出窗体时，在窗体的显示区域的左上角将显示一个消息。当拖动鼠标时，跟随着被拖动的鼠标，一个“*”字符将被显示。在这里注意两个变量——x 和 y，它们存放着在鼠标的按下、释放或拖动事件发生时鼠标的位置。接下来，这些坐标将在 paint()方法中被使用，以便在这些事件发生的点显示输出。

例 4-2 处理鼠标事件。

```
import java.awt.*;
import java.awt.event.*;
import javax.swing.*;
public class MouseEventDemo extends JFrame implements
 MouseListener,MouseMotionListener
{
    private int x=-10;
    private int y=-10;
    private String msg=null;
    private String inOrOut=null;
    public MouseEventDemo(String title)
    {
        super(title);
        this.addMouseListener(this);
        this.addMouseMotionListener(this);
        this.setSize(400,300);
        this.setVisible(true);
        this.setDefaultCloseOperation(JFrame.EXIT_ON_CLOSE);
    }
    public void paint(Graphics g)
    {
        super.paint(g);
        g.drawString("x="+x+" ,y="+y,60,60);
```

```
        if(msg!=null)
        {
            g.drawString(msg,x,y);
            msg=null;
        }
        if(inOrOut!=null)
        {
            g.drawString(inOrOut,10,40);
            inOrOut=null;
        }
    }
    public void mousePressed(MouseEvent e)
    {
        x=e.getX();
        y=e.getY();
        msg="按下";
        repaint();
    }
    public void mouseReleased(MouseEvent e)
    {
        x=e.getX();
        y=e.getY();
        msg="释放";
        repaint();
    }
    public void mouseEntered(MouseEvent e)
    {
        inOrOut="进入";
        repaint();
    }
    public void mouseExited(MouseEvent e)
    {
        inOrOut="出去";
        repaint();
    }
    public void mouseClicked(MouseEvent e)
    {
        x=e.getX();
        y=e.getY();
        msg="单击";
        repaint();
    }
    public void mouseMoved(MouseEvent e)
    {
        x=e.getX();
        y=e.getY();
        repaint();
    }
    public void mouseDragged(MouseEvent e)
    {
        x=e.getX();
        y=e.getY();
        msg="*";
        repaint();
    }
```

```
        public static void main(String[] args)
        {
            new MouseEventDemo("鼠标事件演示");
        }
    }
```

程序运行结果如图 4.5 所示。

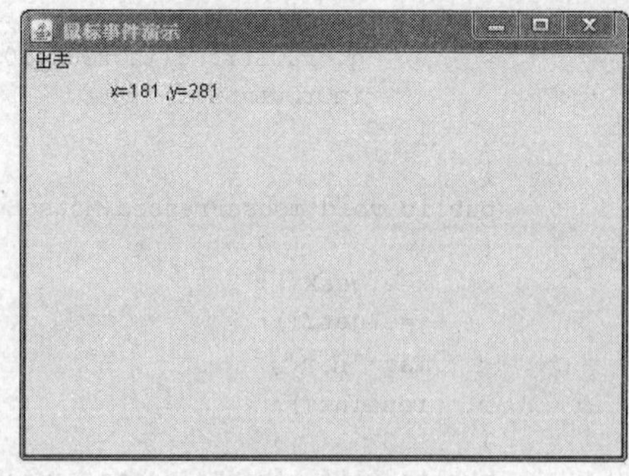

图 4.5　运行结果

MouseEventDemo 类扩展 JFrame 类，同时实现了 MouseListener 接口和 MouseMotionListener 接口。这两个接口包括了接收并处理各种鼠标事件的方法。请注意，这里 JFrame 程序不但是事件源，同时也是这些事件的监听器。这是因为支持 addMouseListener()方法和 addMouseMotionListener()方法的 Component 类是 JFrame 的超类，所以 applet 程序不但是事件源而且是监听器。

在构造方法中，这个程序注册自己为鼠标事件的监听器。这些是通过调用 addMouseListener()方法和 addMouseMotionListener()方法来实现的，它们是类的成员方法，其原型如下：

```
synchronized void addMouseListener(MouseListener ml)
synchronized void addMouseMotionListener(MouseMotionListener mml)
```

这里 ml 是一个接收鼠标事件的对象的引用，而 mml 是一个接收鼠标运动事件的对象的引用。在这个程序里，它们是相同的一个对象。

接下来，这个程序实现了在 MouseListener 接口和 MouseMotionListener 接口中定义的所有方法，以便对这些鼠标事件进行处理。每一种方法都处理了相应的事件，然后返回。

我们可以采用与在前面鼠标事件范例相同的结构去处理键盘事件。当然，必须实现相应的 KeyListener 接口。在分析这个例子之前，让我们回顾一下键盘事件是如何产生的。当一个键被按下时，一个 KEY_PRESSED 事件被产生，这就使 keyPressed()这个事件处理方法被调用。当这个键被释放时，一个 KEY_RELEASED 事件产生，相应的事件处理方法 keyReleased()被执行。如果一个字符被按键产生，那么一个 KEY_TYPED 事件将被产生，并且事件处理方法 keyTyped()将被调用。因此，每次用户按下一个键时，通常有 2 个或 3 个事件被产生。如果只关心字符，那么可以忽略由按键和释放键所产生的信息。然而，如果程序需要处理特殊的键，如方向键，那么就必须通过调用 keyPressed()这个事件处理方法来处理它们。

另外，在程序处理键盘事件之前，必须要获得输入焦点。通过调用 requestFocus()方法可以获得焦点，这种方法在 Component 类中被定义。初学者容易忽略了这一点，若没有获得输入焦点，程序将不会获得任何键盘事件。

下面的程序演示了键盘输入的处理。它将回显按键到程序窗口，并在窗体上显示每一个按键被按下或释放的状态。

例 4-3　处理键盘事件。

```
import java.awt.*;
import java.awt.event.*;
import javax.swing.*;
public class KeyEventDemo extends JFrame implements KeyListener
{
```

```
    private String msg=null;
    private char ch;
    public KeyEventDemo(String title)
    {
        super(title);
        this.requestFocus(true);
        this.addKeyListener(this);
        this.setSize(300,200);
        this.setVisible(true);
        this.setDefaultCloseOperation(JFrame.EXIT_ON_CLOSE);
    }
    public void paint(Graphics g)
    {
        super.paint(g);
        if(msg!=null)
        {
             g.drawString(msg,60,60);
           msg=null;
        }
        g.drawString(String.valueOf(ch),100,100);
    }
    public void keyPressed(KeyEvent e)
    {
        msg="按下";
        repaint();
    }
    public void keyReleased(KeyEvent e)
    {
        msg="释放";
        repaint();
    }
    public void keyTyped(KeyEvent e)
    {
        ch=e.getKeyChar();
        repaint();
    }
    public static void main(String[] args)
    {
        new KeyEventDemo("键盘事件演示");
    }
}
```

程序运行结果如图 4.6 所示。

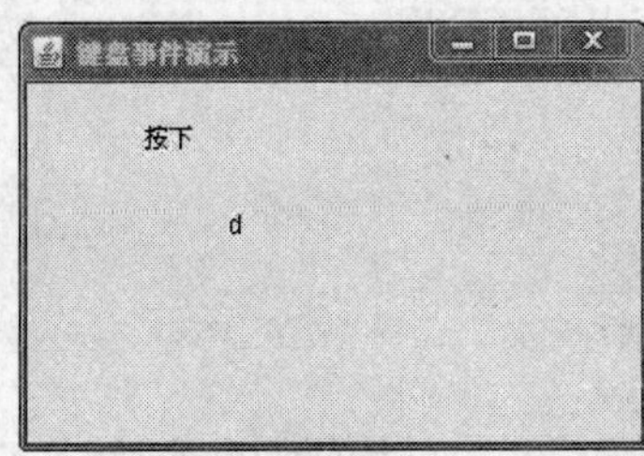

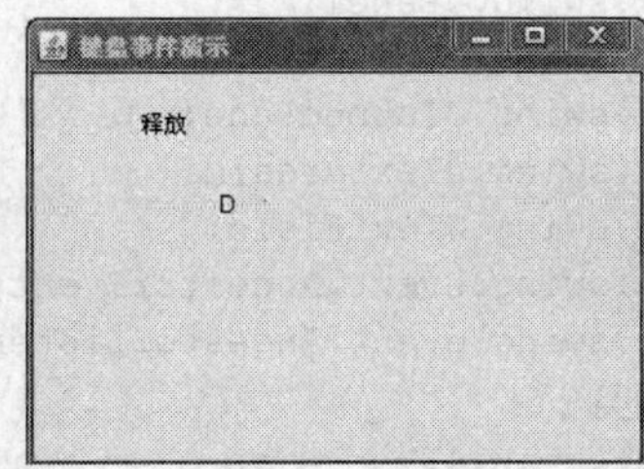

图 4.6　运行结果

如果想处理特殊的键，如方向键，需要在 keyPressed()这个事件处理方法中进行处理。它们不能通过 keyTyped()这个事件处理方法来处理。为了表示这些键，需要使用它们的虚拟键值。

在键盘和鼠标事件处理的例子中展示的程序过程，可以被用于任何类型的事件处理，包括那些由控件产生的事件。

4.5 界面编程实例

针对上述内容，给出一个具体的实例，对公司通讯录系统的主要界面进行编程实现。

```
package 公司通信录;
public class Main {
    public static void main(String args[]) {
        try {
            MainFrame frame = new MainFrame();  //主窗口对象
            frame.setVisible(true); //设置窗口可见
        } catch (Exception e) {
            e.printStackTrace();
        }
    }
}

package 公司通信录;
import java.awt.BorderLayout;
import java.awt.Color;
import java.awt.GridLayout;
import java.awt.event.ActionEvent;
import java.awt.event.ActionListener;
import java.awt.event.MouseAdapter;
import java.awt.event.MouseEvent;
import java.sql.ResultSet;
import java.sql.SQLException;
import java.text.SimpleDateFormat;
import java.util.Calendar;
import javax.swing.ImageIcon;
import javax.swing.JButton;
import javax.swing.JFrame;
import javax.swing.JLabel;
import javax.swing.JMenu;
import javax.swing.JMenuBar;
import javax.swing.JMenuItem;
import javax.swing.JOptionPane;
import javax.swing.JPopupMenu;
import javax.swing.JPanel;
import javax.swing.JScrollPane;
import javax.swing.JTabbedPane;
import javax.swing.JTextArea;
import javax.swing.JTextField;
import javax.swing.event.AncestorEvent;
import javax.swing.event.AncestorListener;
//导入使用到的类
class MainFrame extends  JFrame                             //定义窗口容器对象
{
    private static final long serialVersionUID = 1L;
    DbOperation db = new DbOperation();                     //生成数据库连接对象
    boolean allow = false;
```

```
boolean infisshow=true;
boolean update = false;
String STR;
ResultSet rs;
Object choose;
InfFreme inf;
JTextArea[] textArea = new JTextArea[3];    //生成文本框
JTextField[] textField = new JTextField[100];

public MainFrame() {                         //以下是设置窗口中的对象及布局
    super("公司通信录beta2");
    setJMenuBar(null);
    setBounds(100, 10, 869, 406);
    setDefaultCloseOperation(JFrame.EXIT_ON_CLOSE);

    final JTabbedPane tabbedPane = new JTabbedPane();
    getContentPane().add(tabbedPane, BorderLayout.CENTER);

    final JPanel panel_4 = new JPanel();    //面板对象
    panel_4.setLayout(null);
    tabbedPane.addTab("常用信息", null, panel_4, null);

    final JLabel l_name = new JLabel();     //标签对象
    l_name.setText("姓       名");
    l_name.setBounds(10, 10, 60, 15);
    panel_4.add(l_name);

    final JLabel l_sex = new JLabel();
    l_sex.setText("性       别");
    l_sex.setBounds(10, 31, 60, 15);
    panel_4.add(l_sex);

    final JLabel l_grade = new JLabel();
    l_grade.setText("等       级");
    l_grade.setBounds(10, 52, 60, 15);
    panel_4.add(l_grade);

    final JLabel l_inf_from = new JLabel();
    l_inf_from.setText("信息来源");
    l_inf_from.setBounds(10, 73, 60, 15);
    panel_4.add(l_inf_from);

    final JLabel l_c_name = new JLabel();
    l_c_name.setText("单位名称");
    l_c_name.setBounds(10, 94, 60, 15);
    panel_4.add(l_c_name);

    final JLabel l_branch = new JLabel();
    l_branch.setText("所在部门");
    l_branch.setBounds(10, 115, 60, 15);
    panel_4.add(l_branch);

    final JLabel l_c_addr = new JLabel();
    l_c_addr.setText("单位地址");
```

```
l_c_addr.setBounds(10, 136, 60, 15);
panel_4.add(l_c_addr);

final JLabel l_c_addrs = new JLabel();
l_c_addrs.setText("详细地址");
l_c_addrs.setBounds(10, 157, 60, 15);
panel_4.add(l_c_addrs);

final JLabel l_email = new JLabel();
l_email.setText("常用邮箱");
l_email.setBounds(10, 178, 60, 15);
panel_4.add(l_email);

final JLabel l_user_defined1 = new JLabel();
l_user_defined1.setText("自定义 1");
l_user_defined1.setBounds(10, 199, 60, 15);
panel_4.add(l_user_defined1);

final JLabel l_add_date = new JLabel();
l_add_date.setText("登记日期");
l_add_date.setBounds(10, 220, 60, 15);
panel_4.add(l_add_date);

textField[1] = new JTextField("1");
textField[1].setBounds(66, 8, 140, 19);
panel_4.add(textField[1]);

textField[2] = new JTextField("2");
textField[2].setBounds(66, 29, 140, 19);
panel_4.add(textField[2]);

textField[3] = new JTextField("3");
textField[3].setBounds(66, 50, 140, 19);
panel_4.add(textField[3]);

textField[4] = new JTextField("4");
textField[4].setBounds(66, 71, 140, 19);
panel_4.add(textField[4]);

textField[5] = new JTextField("5");
textField[5].setBounds(66, 92, 363, 19);
panel_4.add(textField[5]);

textField[6] = new JTextField("6");
textField[6].setBounds(66, 113, 140, 19);
panel_4.add(textField[6]);

textField[7] = new JTextField("7");
textField[7].setBounds(66, 134, 363, 19);
panel_4.add(textField[7]);

textField[8] = new JTextField("8");
textField[8].setBounds(66, 155, 363, 19);
panel_4.add(textField[8]);

textField[9] = new JTextField("9");
```

```
textField[9].setBounds(66, 176, 363, 19);
panel_4.add(textField[9]);

textField[10] = new JTextField("10");
textField[10].setBounds(66, 197, 363, 19);
panel_4.add(textField[10]);

textField[11] = new JTextField("11");
textField[11].setBounds(66, 218, 140, 19);
panel_4.add(textField[11]);

final JLabel l_mobile_phone = new JLabel();
l_mobile_phone.setText("手      机");
l_mobile_phone.setBounds(212, 10, 60, 15);
panel_4.add(l_mobile_phone);

final JLabel l_phone_work = new JLabel();
l_phone_work.setText("办公电话");
l_phone_work.setBounds(212, 31, 60, 15);
panel_4.add(l_phone_work);

final JLabel l_fax = new JLabel();
l_fax.setText("办公传真");
l_fax.setBounds(212, 52, 60, 15);
panel_4.add(l_fax);

final JLabel l_XLT_phone = new JLabel();
l_XLT_phone.setText("小  灵  通");
l_XLT_phone.setBounds(212, 73, 60, 15);
panel_4.add(l_XLT_phone);

final JLabel l_duty = new JLabel();
l_duty.setText("所任职务");
l_duty.setBounds(212, 115, 60, 15);
panel_4.add(l_duty);

final JLabel l_group = new JLabel();
l_group.setText("隶属分组");
l_group.setBounds(212, 220, 60, 15);
panel_4.add(l_group);

textField[12] = new JTextField("12");
textField[12].setBounds(268, 8, 161, 19);
panel_4.add(textField[12]);

textField[13] = new JTextField("13");
textField[13].setBounds(268, 29, 161, 19);
panel_4.add(textField[13]);

textField[14] = new JTextField("14");
textField[14].setBounds(268, 50, 161, 19);
panel_4.add(textField[14]);

textField[15] = new JTextField("15");
textField[15].setBounds(268, 71, 161, 19);
panel_4.add(textField[15]);
```

```
textField[16] = new JTextField("16");
textField[16].setBounds(268, 113, 161, 19);
panel_4.add(textField[16]);

textField[17] = new JTextField("17");
textField[17].setBounds(268, 218, 161, 19);
panel_4.add(textField[17]);

Calendar cal = Calendar.getInstance();                    //生成日历对象
java.text.SimpleDateFormat sdf = new SimpleDateFormat(
        "yyyy-MM-dd HH:mm:ss");
String cdate = sdf.format(cal.getTime());

final JLabel label_16 = new JLabel();
label_16.setText("更新日期: " + cdate);
label_16.setBounds(443, 220, 220, 15);
panel_4.add(label_16);

final JScrollPane scrollPane = new JScrollPane();         //生成滚动条
scrollPane.setBounds(443, 10, 403, 204);
panel_4.add(scrollPane);

textArea[0] = new JTextArea();
scrollPane.setViewportView(textArea[0]);

final JPanel panel_4_1 = new JPanel();
panel_4_1.setLayout(null);
panel_4_1.setInheritsPopupMenu(true);
panel_4_1.setFocusTraversalPolicyProvider(true);
panel_4_1.setFocusCycleRoot(true);
panel_4_1.setAutoscrolls(true);
tabbedPane.addTab("个人信息", null, panel_4_1, null);

final JLabel label_17 = new JLabel();
label_17.setText("姓        名");
label_17.setBounds(10, 10, 60, 15);
panel_4_1.add(label_17);

final JLabel label_1_1 = new JLabel();
label_1_1.setText("性        别");
label_1_1.setBounds(10, 31, 60, 15);
panel_4_1.add(label_1_1);

final JLabel label_2_1 = new JLabel();
label_2_1.setText("爱好特长");
label_2_1.setBounds(10, 52, 60, 15);
panel_4_1.add(label_2_1);

final JLabel label_3_1 = new JLabel();
label_3_1.setText("小  灵  通");
label_3_1.setBounds(10, 73, 60, 15);
panel_4_1.add(label_3_1);

final JLabel label_4_1 = new JLabel();
```

```
label_4_1.setText("纪  念  日");
label_4_1.setBounds(10, 94, 60, 15);
panel_4_1.add(label_4_1);

final JLabel label_5_1 = new JLabel();
label_5_1.setText("家庭地址");
label_5_1.setBounds(10, 115, 60, 15);
panel_4_1.add(label_5_1);

final JLabel label_6_1 = new JLabel();
label_6_1.setText("详细地址");
label_6_1.setBounds(10, 136, 60, 15);
panel_4_1.add(label_6_1);

final JLabel label_7_1 = new JLabel();
label_7_1.setText("网络名称");
label_7_1.setBounds(10, 157, 60, 15);
panel_4_1.add(label_7_1);

final JLabel label_8_1 = new JLabel();
label_8_1.setText("常用邮箱");
label_8_1.setBounds(10, 178, 60, 15);
panel_4_1.add(label_8_1);

final JLabel label_9_1 = new JLabel();
label_9_1.setText("备用邮箱");
label_9_1.setBounds(10, 199, 60, 15);
panel_4_1.add(label_9_1);

final JLabel mSNLabel = new JLabel();
mSNLabel.setText("M    S    N");
mSNLabel.setBounds(10, 220, 60, 15);
panel_4_1.add(mSNLabel);

textField[18] = new JTextField("18");
textField[18].setBounds(66, 8, 140, 19);
panel_4_1.add(textField[18]);

textField[19] = new JTextField("19");
textField[19].setBounds(66, 29, 140, 19);
panel_4_1.add(textField[19]);

textField[20] = new JTextField("20");
textField[20].setBounds(66, 50, 140, 19);
panel_4_1.add(textField[20]);

textField[21] = new JTextField("21");
textField[21].setBounds(66, 71, 140, 19);
panel_4_1.add(textField[21]);

textField[22] = new JTextField("22");
textField[22].setBounds(66, 92, 140, 19);
panel_4_1.add(textField[22]);

textField[23] = new JTextField("23");
textField[23].setBounds(66, 113, 363, 19);
```

```
panel_4_1.add(textField[23]);

textField[24] = new JTextField("24");
textField[24].setBounds(66, 134, 363, 19);
panel_4_1.add(textField[24]);

textField[25] = new JTextField("25");
textField[25].setBounds(66, 155, 89, 19);
panel_4_1.add(textField[25]);

textField[26] = new JTextField("26");
textField[26].setBounds(66, 176, 363, 19);
panel_4_1.add(textField[26]);

textField[27] = new JTextField("27");
textField[27].setBounds(66, 197, 363, 19);
panel_4_1.add(textField[27]);

textField[28] = new JTextField("28");
textField[28].setBounds(66, 218, 363, 19);
panel_4_1.add(textField[28]);

final JLabel label_11_1 = new JLabel();
label_11_1.setText("常用手机");
label_11_1.setBounds(212, 10, 60, 15);
panel_4_1.add(label_11_1);

final JLabel label_12_1 = new JLabel();
label_12_1.setText("备用手机");
label_12_1.setBounds(212, 31, 60, 15);
panel_4_1.add(label_12_1);

final JLabel label_13_1 = new JLabel();
label_13_1.setText("家庭电话");
label_13_1.setBounds(212, 52, 60, 15);
panel_4_1.add(label_13_1);

final JLabel label_14_1 = new JLabel();
label_14_1.setText("家庭传真");
label_14_1.setBounds(212, 73, 60, 15);
panel_4_1.add(label_14_1);

textField[29] = new JTextField("29");
textField[29].setBounds(268, 8, 161, 19);
panel_4_1.add(textField[29]);

textField[30] = new JTextField("30");
textField[30].setBounds(268, 29, 161, 19);
panel_4_1.add(textField[30]);

textField[31] = new JTextField("31");
textField[31].setBounds(268, 50, 161, 19);
panel_4_1.add(textField[31]);

textField[32] = new JTextField("32");
textField[32].setBounds(268, 71, 161, 19);
```

```
panel_4_1.add(textField[32]);

final JLabel label_14_1_1 = new JLabel();
label_14_1_1.setText("生        日");
label_14_1_1.setBounds(212, 94, 60, 15);
panel_4_1.add(label_14_1_1);

textField[33] = new JTextField("33");
textField[33].setBounds(268, 92, 161, 19);
panel_4_1.add(textField[33]);

final JLabel qqLabel = new JLabel();
qqLabel.setText("QQ");
qqLabel.setBounds(161, 157, 18, 15);
panel_4_1.add(qqLabel);

final JLabel skypeLabel = new JLabel();
skypeLabel.setText("Skype");
skypeLabel.setBounds(278, 157, 60, 15);
panel_4_1.add(skypeLabel);

textField[34] = new JTextField("34");
textField[34].setBounds(185, 154, 87, 19);
panel_4_1.add(textField[34]);

textField[35] = new JTextField("35");
textField[35].setBounds(319, 154, 111, 19);
panel_4_1.add(textField[35]);

final JLabel label_17_1 = new JLabel();
label_17_1.setText("政治面貌");
label_17_1.setBounds(435, 8, 60, 15);
panel_4_1.add(label_17_1);

textField[36] = new JTextField("36");
textField[36].setBounds(491, 6, 140, 19);
panel_4_1.add(textField[36]);

textField[37] = new JTextField("37");
textField[37].setBounds(491, 27, 140, 19);
panel_4_1.add(textField[37]);

final JLabel label_1_1_1 = new JLabel();
label_1_1_1.setText("技术职称");
label_1_1_1.setBounds(435, 29, 60, 15);
panel_4_1.add(label_1_1_1);

final JLabel label_2_1_1 = new JLabel();
label_2_1_1.setText("最高学历");
label_2_1_1.setBounds(435, 50, 60, 15);
panel_4_1.add(label_2_1_1);

textField[38] = new JTextField("38");
textField[38].setBounds(491, 48, 140, 19);
panel_4_1.add(textField[38]);
```

```
textField[39] = new JTextField("39");
textField[39].setBounds(491, 69, 140, 19);
panel_4_1.add(textField[39]);

final JLabel label_3_1_1 = new JLabel();
label_3_1_1.setText("所学专业");
label_3_1_1.setBounds(435, 71, 60, 15);
panel_4_1.add(label_3_1_1);

final JLabel label_4_1_1 = new JLabel();
label_4_1_1.setText("毕业院校");
label_4_1_1.setBounds(435, 92, 60, 15);
panel_4_1.add(label_4_1_1);

textField[40] = new JTextField("40");
textField[40].setBounds(491, 90, 140, 19);
panel_4_1.add(textField[40]);

final JLabel label_6_1_1 = new JLabel();
label_6_1_1.setText("证件号码");
label_6_1_1.setBounds(435, 134, 60, 15);
panel_4_1.add(label_6_1_1);

textField[41] = new JTextField("41");
textField[41].setBounds(491, 132, 363, 19);
panel_4_1.add(textField[41]);

textField[42] = new JTextField("42");
textField[42].setBounds(491, 174, 363, 19);
panel_4_1.add(textField[42]);

final JLabel label_8_1_1 = new JLabel();
label_8_1_1.setText("纪念摘要");
label_8_1_1.setBounds(435, 176, 60, 15);
panel_4_1.add(label_8_1_1);

final JLabel label_9_1_1 = new JLabel();
label_9_1_1.setText("自定义 9");
label_9_1_1.setBounds(435, 197, 60, 15);
panel_4_1.add(label_9_1_1);

textField[43] = new JTextField("43");
textField[43].setBounds(491, 195, 363, 19);
panel_4_1.add(textField[43]);

textField[44] = new JTextField("44");
textField[44].setBounds(491, 216, 363, 19);
panel_4_1.add(textField[44]);

final JLabel mSNLabel_1 = new JLabel();
mSNLabel_1.setText("自定义 10");
mSNLabel_1.setBounds(435, 220, 60, 15);
panel_4_1.add(mSNLabel_1);

final JLabel label_5_1_1 = new JLabel();
label_5_1_1.setText("个人主页");
```

```
label_5_1_1.setBounds(436, 155, 60, 15);
panel_4_1.add(label_5_1_1);

textField[45] = new JTextField("45");
textField[45].setBounds(492, 153, 363, 19);
panel_4_1.add(textField[45]);

final JLabel label_4_1_1_1 = new JLabel();
label_4_1_1_1.setText("证件类型");
label_4_1_1_1.setBounds(435, 113, 60, 15);
panel_4_1.add(label_4_1_1_1);

textField[46] = new JTextField("46");
textField[46].setBounds(491, 111, 140, 19);
panel_4_1.add(textField[46]);

final JScrollPane scrollPane_2 = new JScrollPane();
scrollPane_2.setBounds(637, 6, 217, 124);
panel_4_1.add(scrollPane_2);

textArea[1] = new JTextArea();
scrollPane_2.setViewportView(textArea[1]);

panel_4_1.addAncestorListener(new AncestorListener() {  //注册事件监听对象
    public void ancestorAdded(AncestorEvent event) {
        textField[18].setText(textField[1].getText()); //设置文本框内容
        textField[19].setText(textField[2].getText());
        textField[21].setText(textField[15].getText());
        textField[26].setText(textField[9].getText());
        textField[29].setText(textField[12].getText());
    }
    public void ancestorMoved(AncestorEvent event) {  // 空方法
    }
    public void ancestorRemoved(AncestorEvent event) { // 空方法

    }
});

final JPanel panel_2 = new JPanel();
panel_2.setLayout(null);
tabbedPane.addTab("单位信息", null, panel_2, null);

final JLabel label_17_2 = new JLabel();
label_17_2.setText("单位名称");
label_17_2.setBounds(10, 11, 60, 15);
panel_2.add(label_17_2);

textField[47] = new JTextField("47");
textField[47].setBounds(66, 8, 363, 19);
panel_2.add(textField[47]);

textField[48] = new JTextField("48");
textField[48].setBounds(66, 32, 140, 19);
panel_2.add(textField[48]);

final JLabel label_1_1_2 = new JLabel();
```

```
label_1_1_2.setText("所在部门");
label_1_1_2.setBounds(10, 35, 60, 15);
panel_2.add(label_1_1_2);

final JLabel label_2_1_2 = new JLabel();
label_2_1_2.setText("单位性质");
label_2_1_2.setBounds(10, 60, 60, 15);
panel_2.add(label_2_1_2);

textField[49] = new JTextField("49");
textField[49].setBounds(66, 57, 140, 19);
panel_2.add(textField[49]);

textField[50] = new JTextField("50");
textField[50].setBounds(66, 81, 363, 19);
panel_2.add(textField[50]);

final JLabel label_3_1_2 = new JLabel();
label_3_1_2.setText("主营业务");
label_3_1_2.setBounds(10, 81, 60, 15);
panel_2.add(label_3_1_2);

final JLabel label_4_1_2 = new JLabel();
label_4_1_2.setText("开户银行");
label_4_1_2.setBounds(10, 106, 60, 15);
panel_2.add(label_4_1_2);

textField[51] = new JTextField("51");
textField[51].setBounds(66, 106, 363, 19);
panel_2.add(textField[51]);

textField[52] = new JTextField("52");
textField[52].setBounds(66, 131, 363, 19);
panel_2.add(textField[52]);

final JLabel label_5_1_2 = new JLabel();
label_5_1_2.setText("单位账号");
label_5_1_2.setBounds(10, 134, 60, 15);
panel_2.add(label_5_1_2);

final JLabel label_6_1_2 = new JLabel();
label_6_1_2.setText("单位税号");
label_6_1_2.setBounds(10, 156, 60, 15);
panel_2.add(label_6_1_2);

textField[53] = new JTextField("53");
textField[53].setBounds(66, 153, 363, 19);
panel_2.add(textField[53]);

textField[54] = new JTextField("54");
textField[54].setBounds(66, 178, 363, 19);
panel_2.add(textField[54]);

final JLabel label_7_1_1 = new JLabel();
label_7_1_1.setText("单位地址");
```

```
label_7_1_1.setBounds(10, 181, 60, 15);
panel_2.add(label_7_1_1);

final JLabel label_8_1_2 = new JLabel();
label_8_1_2.setText("详细地址");
label_8_1_2.setBounds(10, 201, 60, 15);
panel_2.add(label_8_1_2);

textField[55] = new JTextField("55");
textField[55].setBounds(66, 200, 363, 19);
panel_2.add(textField[55]);

textField[56] = new JTextField("56");
textField[56].setBounds(66, 222, 363, 19);
panel_2.add(textField[56]);

final JLabel label_9_1_2 = new JLabel();
label_9_1_2.setText("单位网址");
label_9_1_2.setBounds(10, 225, 60, 15);
panel_2.add(label_9_1_2);

final JLabel label_17_2_1 = new JLabel();
label_17_2_1.setText("所任职务");
label_17_2_1.setBounds(212, 32, 60, 15);
panel_2.add(label_17_2_1);

textField[57] = new JTextField("57");
textField[57].setBounds(268, 29, 161, 19);
panel_2.add(textField[57]);

textField[58] = new JTextField("58");
textField[58].setBounds(268, 53, 161, 19);
panel_2.add(textField[58]);

final JLabel label_1_1_2_1 = new JLabel();
label_1_1_2_1.setText("所在行业");
label_1_1_2_1.setBounds(212, 56, 60, 15);
panel_2.add(label_1_1_2_1);

final JLabel label_17_2_2 = new JLabel();
label_17_2_2.setText("办公电话");
label_17_2_2.setBounds(435, 11, 60, 15);
panel_2.add(label_17_2_2);

textField[59] = new JTextField("59");
textField[59].setBounds(491, 8, 151, 19);
panel_2.add(textField[59]);

textField[60] = new JTextField("60");
textField[60].setBounds(491, 32, 151, 19);
panel_2.add(textField[60]);

final JLabel label_1_1_2_2 = new JLabel();
label_1_1_2_2.setText("办公传真");
label_1_1_2_2.setBounds(435, 35, 60, 15);
panel_2.add(label_1_1_2_2);
```

```
final JLabel label_17_2_2_1 = new JLabel();
label_17_2_2_1.setText("自定义 1");
label_17_2_2_1.setBounds(435, 57, 60, 15);
panel_2.add(label_17_2_2_1);

final JLabel label_1_1_2_2_1 = new JLabel();
label_1_1_2_2_1.setText("自定义 2");
label_1_1_2_2_1.setBounds(435, 84, 60, 15);
panel_2.add(label_1_1_2_2_1);

textField[61] = new JTextField("61");
textField[61].setBounds(491, 81, 151, 19);
panel_2.add(textField[61]);

textField[62] = new JTextField("62");
textField[62].setBounds(491, 57, 151, 19);
panel_2.add(textField[62]);

textField[63] = new JTextField("63");
textField[63].setBounds(491, 178, 151, 19);
panel_2.add(textField[63]);

final JLabel label_1_1_2_2_1_1 = new JLabel();
label_1_1_2_2_1_1.setText("自定义 6");
label_1_1_2_2_1_1.setBounds(435, 181, 60, 15);
panel_2.add(label_1_1_2_2_1_1);

final JLabel label_17_2_2_1_1 = new JLabel();
label_17_2_2_1_1.setText("自定义 5");
label_17_2_2_1_1.setBounds(435, 154, 60, 15);
panel_2.add(label_17_2_2_1_1);

textField[64] = new JTextField("64");
textField[64].setBounds(491, 154, 151, 19);
panel_2.add(textField[64]);

textField[65] = new JTextField("65");
textField[65].setBounds(491, 129, 151, 19);
panel_2.add(textField[65]);

final JLabel label_1_1_2_2_2 = new JLabel();
label_1_1_2_2_2.setText("自定义 4");
label_1_1_2_2_2.setBounds(435, 132, 60, 15);
panel_2.add(label_1_1_2_2_2);

final JLabel label_17_2_2_2 = new JLabel();
label_17_2_2_2.setText("自定义 3");
label_17_2_2_2.setBounds(435, 108, 60, 15);
panel_2.add(label_17_2_2_2);

textField[66] = new JTextField("66");
textField[66].setBounds(491, 105, 151, 19);
panel_2.add(textField[66]);

textField[67] = new JTextField("67");
```

```
textField[67].setBounds(491, 222, 151, 19);
panel_2.add(textField[67]);

final JLabel label_1_1_2_2_1_1_1 = new JLabel();
label_1_1_2_2_1_1_1.setText("自定义 8");
label_1_1_2_2_1_1_1.setBounds(435, 225, 60, 15);
panel_2.add(label_1_1_2_2_1_1_1);

final JLabel label_17_2_2_1_1_1 = new JLabel();
label_17_2_2_1_1_1.setText("自定义 7");
label_17_2_2_1_1_1.setBounds(435, 201, 60, 15);
panel_2.add(label_17_2_2_1_1_1);

textField[68] = new JTextField("68");
textField[68].setBounds(491, 200, 151, 19);
panel_2.add(textField[68]);

final JScrollPane scrollPane_1 = new JScrollPane();
scrollPane_1.setBounds(648, 6, 198, 235);
panel_2.add(scrollPane_1);

textArea[2] = new JTextArea();
scrollPane_1.setViewportView(textArea[2]);
panel_2.addAncestorListener(new AncestorListener() {

    public void ancestorAdded(AncestorEvent event) {

        textField[47].setText(textField[5].getText());
        textField[48].setText(textField[6].getText());
        textField[54].setText(textField[7].getText());
        textField[55].setText(textField[8].getText());
        textField[57].setText(textField[16].getText());
        textField[59].setText(textField[13].getText());
        textField[60].setText(textField[14].getText());
    }
    public void ancestorMoved(AncestorEvent event) {
    }
    public void ancestorRemoved(AncestorEvent event) {
    }
});

final JPanel panel_3 = new JPanel();
panel_3.setLayout(null);
tabbedPane.addTab("名片模式", null, panel_3, null);

final JPanel panel_1 = new JPanel();
panel_1.setLayout(new GridLayout(2, 1));
panel_1.setBounds(280, 10, 356, 210);
panel_3.add(panel_1);

final JPanel panel_6 = new JPanel();
panel_6.setBackground(new Color(95, 158, 160));
panel_1.add(panel_6);

final JPanel panel_5 = new JPanel();
panel_5.setBackground(new Color(176, 224, 230));
panel_5.setLayout(null);
```

```
panel_1.add(panel_5);

final JLabel label_18 = new JLabel();
label_18.setText("公司地址：");
label_18.setBounds(23, 10, 165, 18);
panel_5.add(label_18);

final JLabel label_18_1 = new JLabel();
label_18_1.setText("办公电话：");
label_18_1.setBounds(23, 34, 165, 18);
panel_5.add(label_18_1);

final JLabel label_18_1_1 = new JLabel();
label_18_1_1.setText("手        机：");
label_18_1_1.setBounds(23, 58, 165, 18);
panel_5.add(label_18_1_1);

final JLabel label_18_1_2 = new JLabel();
label_18_1_2.setText("电子邮件：");
label_18_1_2.setBounds(23, 82, 165, 18);
panel_5.add(label_18_1_2);

final JLabel label_18_1_3 = new JLabel();
label_18_1_3.setText("办公传真：");
label_18_1_3.setBounds(194, 34, 152, 18);
panel_5.add(label_18_1_3);

final JLabel label_18_1_4 = new JLabel();
label_18_1_4.setText("公司邮编：");
label_18_1_4.setBounds(194, 58, 152, 18);
panel_5.add(label_18_1_4);
panel_5.addAncestorListener(new AncestorListener() {
    public void ancestorAdded(AncestorEvent event) {
        label_18.setText("公司地址：" + textField[7].getText());
        label_18_1.setText("办公电话：" + textField[13].getText());
        label_18_1_1.setText("手        机：" + textField[12].getText());
        label_18_1_2.setText("电子邮件：" + textField[9].getText());
        label_18_1_3.setText("办公传真：" + textField[14].getText());
        label_18_1_4.setText("公司邮编：");
    }
    public void ancestorMoved(AncestorEvent event) {
    }
    public void ancestorRemoved(AncestorEvent event) {
    }
});

final JPanel panel = new JPanel();
panel.setLayout(new BorderLayout());
tabbedPane.addTab("联系记录", null, panel, null);

PhonePanel ppl=new PhonePanel();
panel.add(ppl.panel2);

final JPanel panel_7 = new JPanel();
```

```
panel_7.setLayout(new BorderLayout());
getContentPane().add(panel_7, BorderLayout.NORTH);

final JPanel panel_8 = new JPanel();
panel_7.add(panel_8, BorderLayout.WEST);

final JButton b_add = new JButton();
panel_8.add(b_add);

b_add.setIcon(new ImageIcon("MainPic/添加.jpg")); // 设置按钮图片
b_add.addActionListener(new ActionListener() {
    public void actionPerformed(ActionEvent e) {
        add();
    }
});

final JButton b_delete = new JButton();
panel_8.add(b_delete);
b_delete.setIcon(new ImageIcon("MainPic/删除.jpg"));
b_delete.addActionListener(new ActionListener() {
    public void actionPerformed(ActionEvent e) {
        del();
    }
});
final JButton b_edit = new JButton();
panel_8.add(b_edit);
b_edit.setIcon(new ImageIcon("MainPic/编辑.jpg"));
b_edit.addActionListener(new ActionListener() {
    public void actionPerformed(ActionEvent e) {
        edit();
    }
});
final JButton b_save = new JButton();
panel_8.add(b_save);
b_save.setIcon(new ImageIcon("MainPic/保存.jpg"));
b_save.addActionListener(new ActionListener() {
    public void actionPerformed(ActionEvent e) {
        save();
    }
});
final JButton b_show = new JButton();
panel_8.add(b_show);
b_show.setIcon(new ImageIcon("MainPic/显示.jpg"));
b_show.addActionListener(new ActionListener() {
    public void actionPerformed(ActionEvent e) {
    infshow();
    }
});

final JButton b_refresh = new JButton();
panel_8.add(b_refresh);
b_refresh.setIcon(new ImageIcon("MainPic/刷新.jpg"));
b_refresh.addActionListener(new ActionListener() {

    public void actionPerformed(ActionEvent e) {
        infrefresh();
```

```
        }
    });

    final JButton b_branch = new JButton();
    panel_8.add(b_branch);
    b_branch.setIcon(new ImageIcon("MainPic/帮助.jpg"));

    final JButton b_exit = new JButton();
    panel_8.add(b_exit);
    b_exit.setIcon(new ImageIcon("MainPic/退出.jpg"));
    b_exit.addActionListener(new ActionListener() {
        public void actionPerformed(ActionEvent event) {
            System.exit(0);
        }
    });

    final JMenuBar menuBar = new JMenuBar();
    JMenu fileMenu = new JMenu("编辑");
    JMenuItem showItem = new JMenuItem("显示");
    showItem.addActionListener(new ActionListener() {
        public void actionPerformed(ActionEvent event) {
            infshow();
        }
    });
    JMenuItem addItem = new JMenuItem("添加");
    addItem.addActionListener(new ActionListener() {
        public void actionPerformed(ActionEvent event) {
            add();
        }
    });
    JMenuItem delItem = new JMenuItem("删除");
    delItem.addActionListener(new ActionListener() {
        public void actionPerformed(ActionEvent event) {
            del();
        }
    });
    JMenuItem editItem = new JMenuItem("编辑");
    editItem.addActionListener(new ActionListener() {
        public void actionPerformed(ActionEvent event) {
            edit();
        }
    });
    JMenuItem saveItem = new JMenuItem("保存");
    saveItem.addActionListener(new ActionListener() {
        public void actionPerformed(ActionEvent event) {
            save();
        }
    });
    JMenuItem exitItem = new JMenuItem("退出");
    exitItem.addActionListener(new ActionListener() {
        public void actionPerformed(ActionEvent event) {
            System.exit(0);
        }
    });
    fileMenu.add(showItem); // 设置菜单
```

```
 fileMenu.add(addItem);
 fileMenu.add(delItem);
 fileMenu.add(editItem);
 fileMenu.add(saveItem);
 fileMenu.add(exitItem);
 JMenu viewMenu = new JMenu("视图");
 final JMenuItem infItem = new JMenuItem("关闭缩略图");
final JMenuItem  item  =  new  JMenuItem("关闭缩略图");
 infItem.addActionListener(new ActionListener() {
     public void actionPerformed(ActionEvent event) {
         if(infisshow)
         {
             inf.setVisible(false);
             item.setText("打开缩略图");
             infItem.setText("打开缩略图");
             infisshow=false;
         }
         else
         {
             inf.setVisible(true);
             item.setText("关闭缩略图");
             infItem.setText("关闭缩略图");
             infisshow=true;
         }
     }
 });
 viewMenu.add(infItem);
 JMenu otherMenu = new JMenu("其他");
 JMenuItem aboutItem = new JMenuItem("关于通讯录");
 otherMenu.add(aboutItem);
 menuBar.add(fileMenu);
 menuBar.add(viewMenu);
 menuBar.add(otherMenu);
 setJMenuBar(menuBar);
 final JPopupMenu popup  =  new  JPopupMenu();  //生成弹出菜单

item.addActionListener(new
        ActionListener()
        {
                public  void  actionPerformed(ActionEvent  e)
                {                                       // 设置按钮方法
                    if(infisshow)
                     {
                          inf.setVisible(false);
                          item.setText("打开缩略图");
                          infItem.setText("打开缩略图");
                          infisshow=false;
                     }
                     else
                     {
                          inf.setVisible(true);
                          item.setText("关闭缩略图");
                          infItem.setText("关闭缩略图");
                          infisshow=true;
```

```
                }
            }
        });
    popup.add(item);

    panel_4.addMouseListener(new
            MouseAdapter()                                    // 鼠标适配器对象
            {
                public  void  mousePressed(MouseEvent  event)
                {                                             // 鼠标按下方法
                    if  (event.isPopupTrigger())
                        popup.show(event.getComponent(),
                                event.getX(),  event.getY());
                }

                public  void  mouseReleased(MouseEvent  event)
                {                                             // 鼠标释放方法
                    if  (event.isPopupTrigger())
                        popup.show(event.getComponent(),
                                event.getX(),  event.getY());
                }
            });
        closeTextField();
        infPanel();

    }

    void add() {
        allow = true;
        openTextField();
    }

    void edit() {
        allow = true;
        update= true;
        openTextField();
        select();

    }
    void save() {                                             // 数据保存在数据库中
        if (allow) {
            if (update) {
                for (int i = 1; i < 69; i++) {
                    db.executeUpdate("update UserInf set [" + i + "]='"
                            + textField[i].getText() + "'where id='"
                            + choose + "'");                  // 更新数据库信息
                }
                db.dbClose();
                JOptionPane.showMessageDialog(null, "保存已成功！", "成功提示",
                        JOptionPane.INFORMATION_MESSAGE);
            } else {
                db.execute("insert into UserInf("+setcolumn()+") values ("
                        + getalltext() + ")");                // 插入数据库信息
                db.dbClose();
                JOptionPane.showMessageDialog(null, "保存已成功！", "成功提示",
                        JOptionPane.INFORMATION_MESSAGE);
```

```
            }
            update = false;
            allow = false;
        } else {
            JOptionPane.showMessageDialog(null, "请确认操作! ", "错误提示",
                    JOptionPane.ERROR_MESSAGE);
        }
        db.dbClose();
        closeTextField();
        infrefresh();


    }

    void del() {
        getchoose();
        if (choose.equals("null")) {
            JOptionPane.showMessageDialog(null, "请选择数据", "错误提示",
                    JOptionPane.ERROR_MESSAGE);
        } else {
            String but[] = { "确定", "取消" };
            JPanel delpanel = new JPanel();
            JLabel dellable = new JLabel("确定删除? ");
            delpanel.add(dellable );
            int go = JOptionPane.showOptionDialog(null, delpanel, "提示信息",
                    JOptionPane.YES_OPTION, JOptionPane.INFORMATION_MESSAGE,
                    null, but, but[0]);
            if (go == 0) {
                db.execute("delete from UserInf where id ='" + choose + "'");
// 删除数据库信息
                JOptionPane.showMessageDialog(null, "删除成功! ", "成功提示",
                        JOptionPane.INFORMATION_MESSAGE);

            }
        }
        db.dbClose();
        infrefresh();


    }

     void infshow(){
         select();
    }

    void infrefresh() {
         inf.setVisible(false);
         infPanel();
    }

    void infPanel() {
         inf = new InfFreme();
         inf.setVisible(true);
         inf.table.getSelectedRow();

    }

    void getchoose() {
```

```
        if(inf.table.getSelectedRow()!=-1)
        {
            choose=inf.body[inf.table.getSelectedRow()][9];      //获取选择的行数据
        }
        else
        {
            choose="null";
        }
    }

    void openTextField() {
        for (int i = 1; i < 69; i++) {
            textField[i].setEditable(true);
            textField[i].setText("");
        }
    }

    void closeTextField() {
        for (int i = 1; i < 69; i++) {
            textField[i].setEditable(false);
            textField[i].setText("");
        }
    }
    void select()
    {
        getchoose();
        if (choose.equals("null")) {

            JOptionPane.showMessageDialog(null, "请选择数据", "错误提示",
                    JOptionPane.ERROR_MESSAGE);
        } else {
            rs = db.executeQuery("select"+setcolumn()+"from UserInf where id='"
                    + choose + "'");         // 查询数据库信息
            try {
                while (rs.next()) {
                    for (int i = 1; i < 69; i++) {
                        textField[i].setText(rs.getString(i));
                    }
                }
            } catch (SQLException e1) {
                e1.printStackTrace();
            }
        }
        db.dbClose();
    }
    String getalltext()
    {
        String temp="'"+textField[1].getText()+"'";
        String temp2=null;
        String temp3=null;
        for(int i=2;i<69;i++)
        {
            temp3=textField[i].getText();
            temp2=temp+",'"+temp3+"'";
            temp=temp2;
        }
        return temp;
```

```
    }

    String setcolumn()
    {
        String temp="[1]";
        String temp2=null;
        int temp3=0;
        for(int i=2;i<69;i++)
        {
            temp3=i;
            temp2=temp+",["+temp3+"]";
            temp=temp2;
        }
        return temp;

    }
}

package 公司通信录;
import java.awt.BorderLayout;
import java.awt.event.WindowAdapter;
import java.awt.event.WindowEvent;
import java.sql.ResultSet;
import java.sql.SQLException;
import javax.swing.JFrame;
import javax.swing.JScrollPane;
import javax.swing.JTabbedPane;
import javax.swing.JTable;
class InfFreme extends JFrame
{
    private static final long serialVersionUID = 1L;
    DbOperation db=new DbOperation();
    ResultSet rs;
    JTable table;
    Object body[][]=new Object[15][11];
    String title[]={"姓名","性别","单位名称","办公电话","办公传真","手机","电子邮件",
                "登记日期","隶属分组","序号"};
    public InfFreme() {
        super();
        setBounds(100, 410, 869, 220);
        addWindowListener(new WindowAdapter(){
            public void windowClosed(WindowEvent e) {
                System.exit(0);                    // 窗口关闭事件处理
            }
        });
        final JTabbedPane tabbedPane = new JTabbedPane();
        getContentPane().add(tabbedPane, BorderLayout.CENTER);
        final JScrollPane scrollPane = new JScrollPane();
        tabbedPane.addTab("信息", null, scrollPane, null);
        table = new JTable(body,title);
        scrollPane.setViewportView(table);
        int i=0;
        rs=db.executeQuery("select [1],[2],[3],[13],[14],[12],[9],[11],[17],id
                                from UserInf");     // 查询数据置入数据组中
        try {
            while(rs.next()){
```

```
                body[i][0]=rs.getString(1);
                body[i][1]=rs.getString(2);
                body[i][2]=rs.getString(3);
                body[i][3]=rs.getString(4);
                body[i][4]=rs.getString(5);
                body[i][5]=rs.getString(6);
                body[i][6]=rs.getString(7);
                body[i][7]=rs.getString(8);
                body[i][8]=rs.getString(9);
                body[i][9]=rs.getString(10);
                i=i+1;
                }
            } catch (SQLException e1) {

                e1.printStackTrace();
            }
            finally{
                db.dbClose();
            }
        }
    }
```

程序运行部分结果如图 4.7～图 4.10 所示。

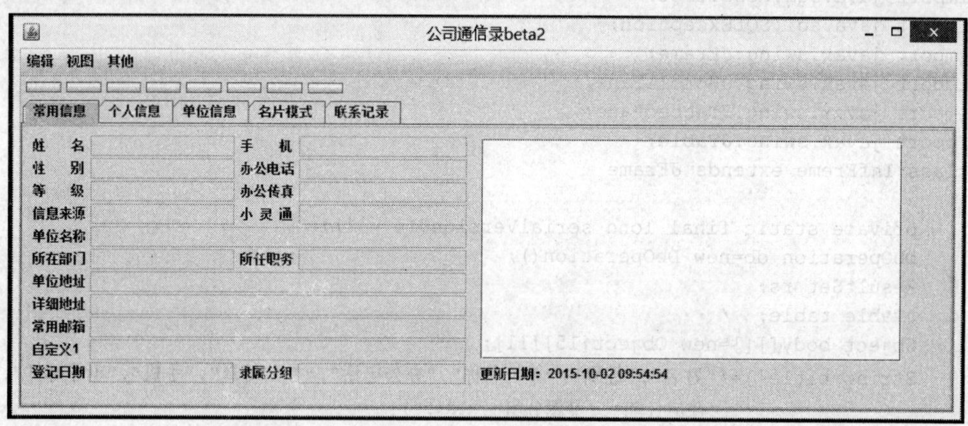

图 4.7 常用信息界面

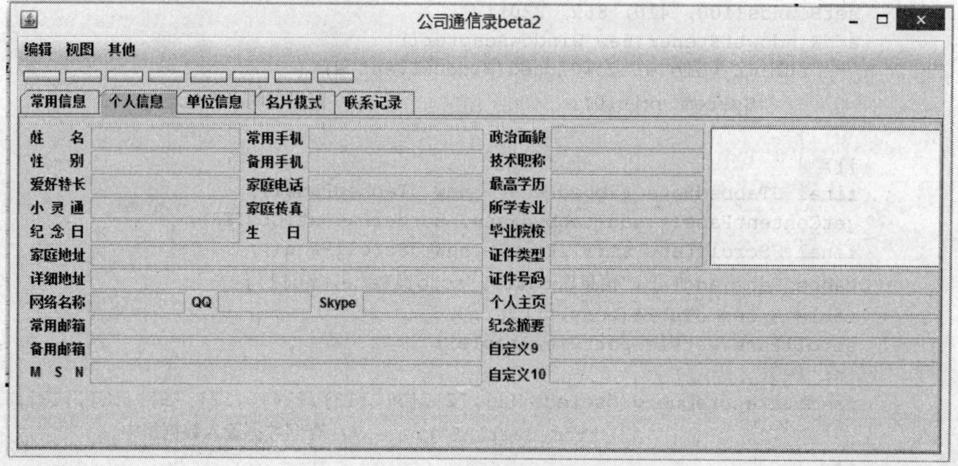

图 4.8 个人信息界面

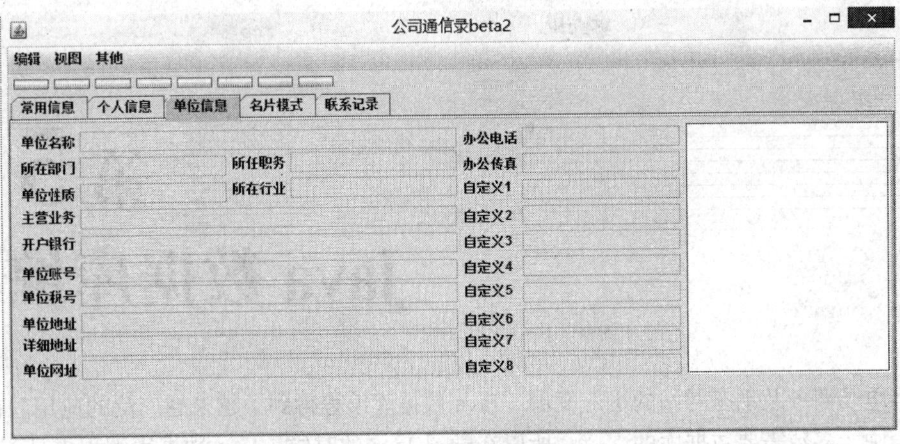

图 4.9　单位信息界面

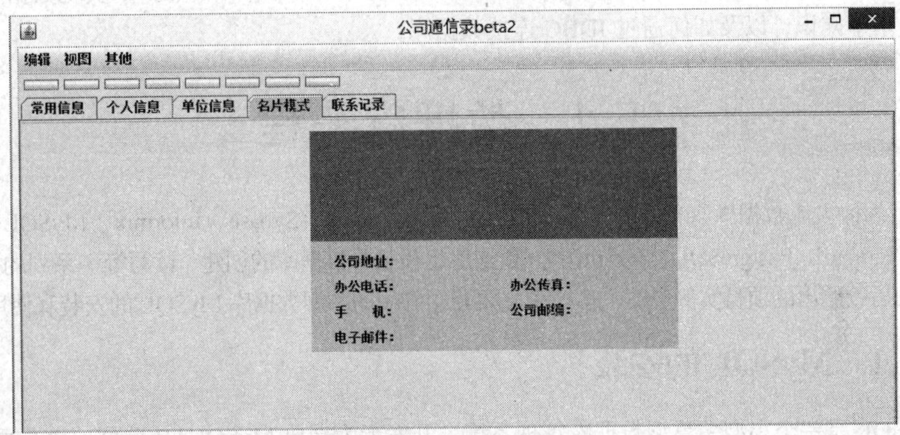

图 4.10　名片模式界面

思考与习题

1. 填空题：图形用户界面或图形用户接口（Graphical User Interface，GUI）是指____________。

2. 填空题：AWT（AbstractWindowToolkit）为抽象窗口工具包，是______________的基本工具。

3. 填空题：在 Swing 组件的 GUI 应用程序中，作为框架用的有 4 个组件，分别是________、________、________和 Jdialog。

4. 简答题：简述利用授权事件模型编写简单的 GUI 应用程序的步骤。

5. 简答题：简述 Java 事件处理模型。

6. 编程题：创建一个自己的密码文本框，输入的数据以“*”字符显示出来。

7. 编程题：创建一个计算器的界面，并实现算术四则运算的功能。

8. 编程题：设计一个简历界面，并添加相应的事件处理程序。

第5章 Java数据库编程

随着电子商务及动态网站的迅速发展，Java数据库编程得到了越来越广泛的应用。在现有的编程语言中，多数需要数据库的支持，所以在学习Java的时候，Java数据库编程也是一个重点。本章介绍了关系型数据库中MySQL数据库的安装和基本知识、JDBC（Java实现数据库访问的API）的基本知识，以及如何通过JDBC连接数据库。

5.1 数据库编程

当前各种主流数据库有很多，包括Oracle、SQL Server、Sybase、Informix、MySQL、DB2、Interbase/Firebird、PostgreSQL、SQLite等。数据库编程是对数据库的创建、读写等一系列的操作。考虑到本书后续给出的项目实例开发，本节中主要介绍常用的小型数据库MySQL的安装和使用。

5.1.1 MySQL的安装

MySQL是一个小型关系型数据库管理系统，开发者为瑞典MySQLAB公司，该公司于2008年1月16日被Sun公司收购。由于其体积小、速度快、总体拥有成本低，尤其是开放源码这一特点，MySQL成为许多中小型网站数据库的首选。在开发Java Web应用时，Tomcat与MySQL的组合是非常流行的服务器端软件的搭配方式。

下面具体介绍如何安装MySQL数据库。

官网http://www.mysql.com中提供了MySQL的安装软件，本书下载MySQL5.5版本作为示例。下载mysql-5.5.19-win32.msi安装文件，用鼠标双击运行，出现图5.1所示界面，单击“Next”按钮继续。

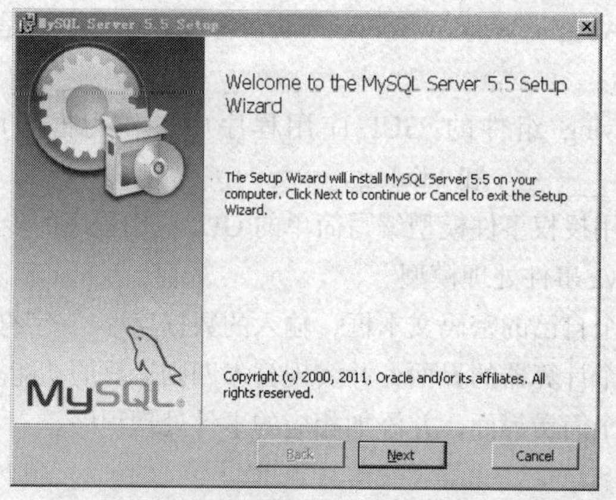

图5.1 MySQL安装向导图

接下来选择安装类型，选择“Custom”选项，单击“Next”按钮继续。在 Custom 自定义安装中，可以选择安装路径，这里将 MySQL 安装到“D:\program files\MySQL”，单击“Install”按钮继续，如图 5.2 所示。

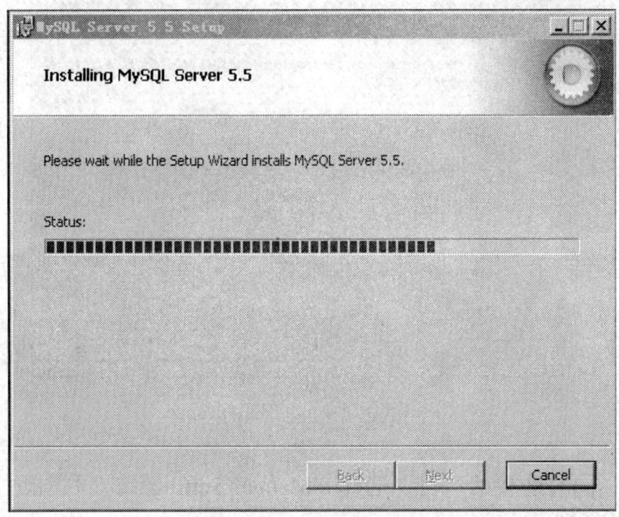

图 5.2　MySQL 安装示意图

根据提示单击“Next”按钮直至出现图 5.3 所示的界面，将“Launch the Mysql Instance Configuration Wizard（启用 MySQL 配置向导）”前面的勾打上，单击“Finish”按钮完成安装。

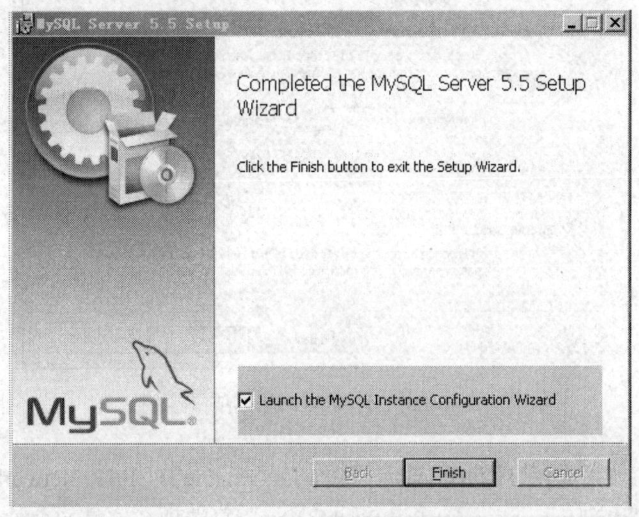

图 5.3　MySQL 配置向导启动图

MySQL 配置向导界面启动，单击“Next”按钮继续，依次按步骤选择“Detailed Configuration”“Developer Machine”“Multifunctional Database”，单击“Next”按钮继续直至如图 5.4 所示。

InnoDB Tablespace 是 InnoDB 数据库文件的存储空间，这里可以使用默认位置，单击“Next”按钮继续。如果修改了，要记住修改的位置，重装的时候选择一样的地方，否则会造成数据库损坏。

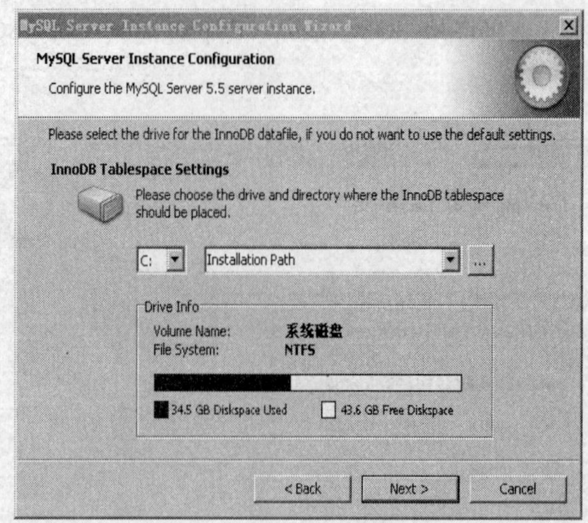

图 5.4　配置 InnoDB Tablespace

接着选择 MySQL 同时连接的数目，选择“Manual Setting”选项，根据自己的实际情况输入数值，单击“Next”按钮继续。

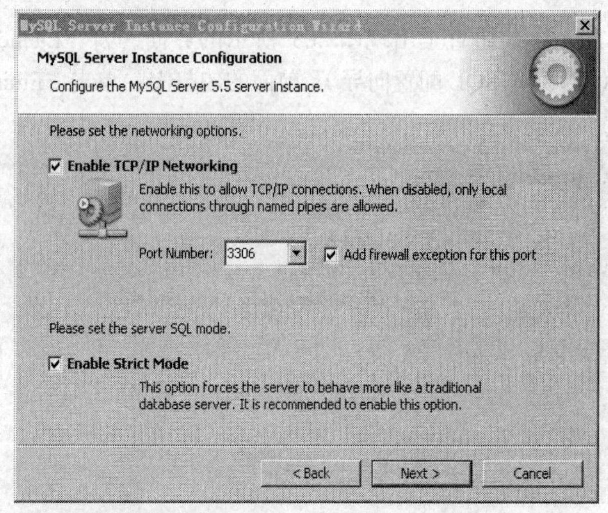

图 5.5　设置网络选项

如图 5.5 所示，如果启用 TCP/IP 连接，则勾上“Enable TCP/IP Networking”复选项，“Port Number”为 3306，“Add firewall exception for this port”复选项，这是一个关于防火墙的设置，将监听端口加为 Windows 防火墙的例外允许端口，避免被防火墙阻断。“Enable Strict Mode”被选中后，MySQL 就不会允许细小的语法错误，建议选择，单击“Next”按钮继续。

设置 MySQL 默认数据库编码时，选择“Manual Selected Default Character Set/Collation”选项，并选择 utf8 编码，单击“Next”按钮继续下一步，如图 5.6 所示。

如图 5.6 所示勾上选项，将 MySQL 安装为 Windows 服务，还可以指定 Service Name（服务标识名称），单击“Next”按钮继续。最后修改默认 root 用户的密码，单击“Next”按钮继续。确保上面的设置无误后，单击“Execute”按钮，使设置生效，如果有误，单击“Back”按钮返回检查。至此安装完毕。

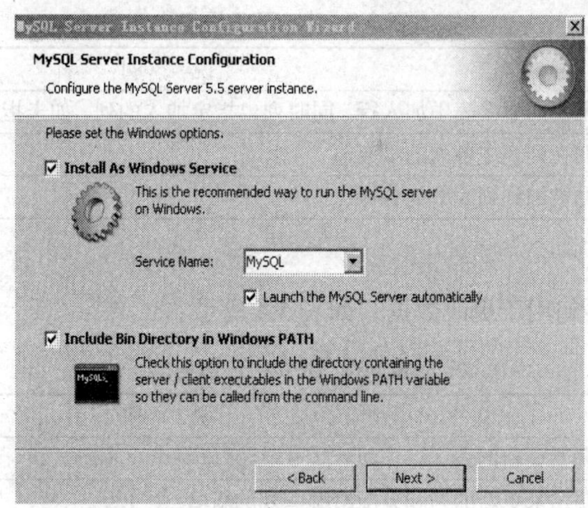

图 5.6　将 MySQL 安装为 Windows 服务

5.1.2　SQL 语言简介

SQL（Structured Query Language）是一种标准化的数据库查询语言。可以用 SQL 实现数据库的定义、增加、修改、删除，以及用户权限的设置等。

SQL 的主要语句如表 5.1 所示。

表 5.1　SQL 的主要语句

命　令	类　别	说　明
SELECT	数据查询语言	检索数据
INSERT	数据操纵语言	向一个表中增加行
UPDATE	数据操纵语言	更新表
DELETE	数据操纵语言	删除行
CREATE	数据定义语言	创建一个新表
DROP	数据定义语言	删除一个表

1. SELECT 语句

SELECT（选择或检索）语句用于从数据库中选择满足条件的记录，如表 5.2 所示。

表 5.2　SELECT 语句

子　句	说　明
SELECT	指明要检索数据的列，如果是“*”，则表示选择所有的列
FROM	指明从哪（儿）个表中进行检索
WHERE	指明返回数据必须满足的条件
GROUP BY	指明返回的列数据通过某些条件来形成组
HAVING	指明返回的集合必须满足的标准
ORDER BY	指明返回的行的排列顺序

2. INSERT 语句

INSERT 语句用于向数据库的某个表中插入新的记录，如表 5.3 所示。

表 5.3　INSERT 语句

子　句	说　明
INSERT INTO	指明要向哪个表中加入行，同时列出指定加入的列，如未指定对象，则为表中的每一列
VALUES	指明在列表中各列的填充值
SELECT	返回被加到列表中的各行

3. UPDATE 语句

UPDATE（更新）语句用于更改数据，如表 5.4 所示。

表 5.4　UPDATE 语句

子　句	说　明
UPDATE	指明要更新的表
SET	指明用来更新的列和分配给哪些列的新值
FROM	指明 UPDATE 语句可以处理的对象表
WHERE	指明要更新的数据所满足的标准

4. DELETE 语句

DELETE 语句用于删除数据表中的指定行，如表 5.5 所示。

表 5.5　DELETE 语句

子　句	说　明
DELETE FROM	指明待删除操作的表
WHERE	指明待删除行所满足的标准

（1）创建数据库 test，SQL 命令如下：

```
CREATE DATABASE test
```

（2）进入 test 数据库，SQL 命令如下：

```
use test
```

（3）在 test 数据库中创建 user 表，SQL 命令如下：

```
CREATE TABLE user
```

（4）删除 test 数据库，SQL 命令如下：

```
DROP DATABASE test
```

5.2　JDBC

JDBC（Java Database Connectivity）是 Java 实现数据库访问的 API（application PRogramming Interface），与 Microsoft 的 ODBC（Open Database Connectivity）一样，JDBC 是建立在 X/Open SQL CLI（Call Level Interface）基础上的。本节中在介绍 JDBC 的基本概念后，介绍其系统编程模型、接口及驱动。

5.2.1　JDBC 概念

JDBC 允许用户从 Java 应用程序中访问任何表格数据源，它由一组用 Java 编程语言编写的类和接口组成。Java API 向 Java 编程语言提供了统一的数据访问方法。使用 JDBC API 可以访问从关系型数据库到表单、一般文件的任何一种数据源。

Java 程序员借助于 JDBC API，能够使用 Java 语言来执行 SQL 语句，从而完成对数据库的 DML 操作；同样也可以在分布式的异质环境下与多个数据源交互。一般来说，JDBC 的工作主要分为 3 个步骤：首先与某一关系数据库建立连接；然后向数据库发送 SQL 语句，实现对数据库的操作；最后取得处理结果。

JDBC 是个“低级”接口，用来支持独立于任何特定 SQL 实现的基本 SQL 功能。JDBC 是以 X/Open SQL CLI（Call Level Interface）为基础的，X/Open SQL CLI 是设计用于访问 SQL 数据库的国际标准，也是 ODBC 的基础。

目前，Microsoft 公司的 ODBC（开放式数据库连接）API 是使用比较广的、用于访问关系数据库的编程接口。它几乎能在所有平台上连接所有的数据库系统。

在 Java 系统实现中，可以使用 ODBC 进行数据源的连接，但不建议这样使用，因为 ODBC 使用 C 语言接口。从 Java 调用本地 C 语言代码在安全性、可实现性、坚固性和程序的自动移植性方面都有许多缺点，而 JDBC API 对于基本的 SQL 抽象和概念是一种自然的 Java 接口，JDBC 保留了 ODBC 的基本设计特征，JDBC 以 Java 风格与优点为基础并进行了优化，因此更易于使用。

JDBC 的主要特点是与任何关系型数据库协同工作的方式完全相同，即只要写一个程序就可以访问 Oracle 数据库、Sybase 数据库和 Informix 数据库，同时由于 Java 语言的跨平台性，只要写一遍程序，就可以在任何平台上运行。

5.2.2　系统编程模型

基于数据库应用系统的编程模型主要有两种：两层模型和三层模型，在有些大型的企业级应用中，还可以有四层模型。

两层模型：多客户端和多数据库服务器架构，即应用层和数据库层。这样的模型被称为 C/S 结构，即客户端/服务器端结构，又称为胖客户机结构。在此模型下，客户端的程序直接与数据库服务器相连接，并发送 SQL 语句（但这时就需要在客户端安装被访问的数据库的 JDBC 驱动程序），DBMS 服务器向客户返回相应的结果，客户程序负责对数据的格式化。

三层模型：在两层模型的基础上再加上一个中间应用服务器，即客户层、中间层和数据库层。客户端往往是由浏览器来充当，这样的模型被称为 B/S 结构，又称瘦客户机结构。在此模型下，客户端将用户的请求发送至应用服务器进行解析，再将数据访问请求转发给数据库服务器，执行后将结果返回给浏览器客户端。

JDBC 同时支持这两种模型。在两层模型中，一个 Java applet 或 Java 应用程序直接与数据库连接。用户的 SQL 语句被传送到数据库，进行数据库处理后，结果返回给用户。其中数据库可以与程序在同一机器上，也可以在另一台机器上，通过网络来进行传输，这个网络可以是 Intranet，也可以是 Internet。用户的计算机作为客户端，运行数据库的计算机作为服务器。在这种模型中，客户端需要能直接访问数据库的 JDBC 驱动器。

主要缺点：受数据库厂商的限制，用户更换数据库时需要改写客户程序；受数据库版本的限制，数据库厂商一旦升级数据库，使用该数据库的客户程序需要重新编译和发布；对数据库的操作与处理都是在客户程序中实现的，使客户程序在编程与设计时较为复杂。

在三层模型中，命令将被发送到服务的中间层，而中间层将 SQL 语句发送到数据库。数据库处理 SQL 语句后将结果返回到中间层，再由中间层将它们返回给客户端。三层模型很有吸引力，因为中间层可以对访问进行控制，并协同数据库的更新。

在此模型下，主要在客户端的程序与数据库服务器之间增加了一个中间服务器（可以采用 C++或 Java 语言来编程实现），隔离客户端的程序与数据库服务器。客户端的程序（可以简单为通用的浏览器）与中间服务器进行通信，然后由中间服务器处理客户端程序的请求，并管理与数据库服务器的连接。

在三层模型结构中，中间件是最重要的组件。所谓中间件是一个用 API 定义的软件层，具有强大的通信能力和良好的可扩展性的分布式软件管理框架。它的功能是：在客户端和服务器或者服务器与服务器之间传送数据，实现客户机群和服务器群之间的通信。

5.2.3 JDBC 接口及驱动

1. JDBC 接口

JDBC 有两种主要接口：面向开发人员的 JDBC 的接口和面向数据库提供商及第三方的接口。

面向开发人员的 JDBC 的接口有 Connection、Statement、ResultSet。Connection 提供 JDBC API 和 URL 指定的数据库系统之间的连接，Statement 对象作为在给定 Connection 上执行 SQL 语句的容器，ResultSet 控制通过游标移动对变化结构中给定 Statement 的结果进行访问。

面向数据库提供商和第三方的接口，是由数据库供应商对上述 3 个接口的实现。这个实现称为数据库驱动程序。例如，OracleConnection 是为 Oracle 数据库系统提供的 Connection 接口的实现，SQLServerConnection 是 SQL Server 2000 数据库系统提供的 Connection 接口的实现，oracle.jdbc.driver.OracleDriver 是 Oracle 所提供的数据库驱动程序类，com.microsoft.jdbc. sqlserver. SQLServerDriver 是微软公司所提供的数据库驱动程序。

2. JDBC 驱动

JDBC 有 3 种产品组件。

（1）JDBC 驱动程序管理器：JDBC 体系结构的支柱。它实际上很小，也很简单；其主要作用是把 Java 应用程序连接到正确的 JDBC 驱动程序上，然后退出。

（2）JDBC 驱动程序测试工具包：给 JDBC 驱动程序运行程序提供一定的可信度。只有通过 JDBC 驱动程序测试包的驱动程序才被认为是符合 JDBC 标准的。

（3）JDBC-ODBC 桥：将 ODBC 驱动程序作为 JDBC 的驱动程序，即利用 ODBC 驱动程序提供 JDBC 访问。

JDBC 驱动程序包含 4 个种类。

（1）JDBC-ODBC 桥和 ODBC 驱动程序。通过 ODBC 驱动，提供 JDBC 存取功能。驱动程序的 java.sql.Drive 接口在 sun.jdbc.odbc.JdbcOdbcDriver 中实现。需要注意的是，必须将 ODBC 二进制代码（许多情况下还包括数据库客户机代码）加载到使用该驱动程序的每个客户机上。因此，这种类型的驱动程序最适合于企业网（这种网络上客户机的安装不是主要问题），或者是用 Java 编写的三层结构的应用程序服务器代码。JDBC-ODBC 桥接方式利用微软的开放数据库互连接口（ODBC API）与数据库服务器通信，客户端计算机首先应该安装并配置 ODBC driver 和 JDBC-ODBC bridge 两种驱动程序。

主要优点是：它可能是访问低端桌面数据库和应用程序的唯一方式；主要缺点是：ODBC 驱动程序必须加载到目标机上，JDBC 和 ODBC 的转换影响性能。

（2）本地 API 部分用 Java 来编写的驱动程序。把 JDBC 调用转换成对客户端数据库管理系统 API 的调用。与 JDBC-ODBC 桥驱动程序相似，这一类型的驱动程序需要在每个客户端上安装和操作系统相关的二进制代码，使 Java 数据库客户端方与数据库服务器方通信。例如，Oracle 用 SQLNet 协议，DB2 用 IBM 的数据库协议。数据库厂商的特殊协议也应该安装在客户机上，但由于这类驱动程序不必经过 ODBC 中间层，因此其性能明显优于由 JDBC-ODBC 桥的方式构成的驱动程序。

（3）JDBC 网络纯 Java 驱动程序。把 JDBC 调用解释为与数据库管理系统独立的网络协议，这种网络协议再被服务器解释成数据库管理系统的协议。这个中间件网络服务器可以把纯 Java 客户端与多个不同的数据库连接起来，所用的协议与生产厂商相关。这是最灵活的 JDBC 可选方案，不需要客户端安装本地库，所有的生产厂商很可能提供适合于内部网使用的产品，要使这些产品支持 Internet，它们必须处理针对安全、防火墙访问等网络方面的额外要求。

这种方式是纯 Java driver。数据库客户以标准网络协议(如 HTTP、SHTTP)同数据库访问服务器通信，数据库访问服务器然后翻译标准网络协议，成为数据库厂商的专有特殊数据库访问协议与数据库通信。对 Internet 和 Intranet 用户而言，这是一个理想的解决方案。Java driver 被自动的、以透明的方式随 Applets 自 Web 服务器而下载，并安装在用户的计算机上。

（4）本地协议纯 Java 驱动程序。直接把 JDBC 调用转换成数据库管理系统使用的网络协议，它允许从客户端机器上直接调用数据库管理系统服务器，这是内部网绝佳的解决方案。这种方式也是纯 Java driver。数据库厂商提供了特殊的 JDBC 协议，使 Java 数据库客户端与数据库服务器通信。由于这些网络协议中很多是厂商专有的，所以客户端配置的灵活性会受到一定的限制。目前，市场上已经有的这类产品包括 Oracel、Sybase、Informix、IBM DB2、Microsoft SQL Server、Inprise InterBase 等。然而，将代理协议同数据库服务器通信改用数据库厂商的特殊 JDBC driver，这对 Intranet 应用是高效的，可是数据库厂商的协议可能不被防火墙支持，缺乏防火墙支持，在 Internet 应用中会存在潜在的安全隐患。驱动程序的比较如表 5.6 所示。

表 5.6　驱动程序的比较

驱动程序类型	纯 Java	网 络 连 接
JDBC-ODBC 桥	否	直接
本地 API 为基础	否	直接
JDBC 网络	客户端是，服务器端可能是	要求连接器
本地协议为基础	是	直接

5.3　通过 JDBC 访问数据库

下面将介绍在 Java 中如何通过 JDBC 访问数据库，并给出实例模型具体说明。

5.3.1　java.sql 包

JDBC 的接口包含在 java.sql 和 javax.sql 包中，其中 java.sql 属于 java SE 平台，javax.sql 属于 java EE 平台，本小节主要对 java.sql 包进行介绍。java.sql 包中定义了很多接口和类，这里介绍 5 个常用的接口和类。

1. 加载驱动程序接口：Driver

java.sql.Driver：Driver 对象实现 acceptsURL（String url）方法，确认自己具有连接到 DriverManager 传递的 URL 上的能力。这是所有 JDBC 驱动程序必须实现的接口。在编程中要连接数据库，必须先装载特定厂家提供的数据库驱动程序（Driver）。例如以下内容。

使用 JDBC-ODBC 桥驱动装载方法：

```
Class.forName= "sun.jdbc.odbc.JdbcOdbcDriver";
```

加载 SQL Server 数据库的驱动：

```
Class.forName("com.microsoft.jdbc.sqlserver.SQLServerDriver");
```

加载 Oracle 数据库的驱动：

```
Class.forName("oracle.jdbc.driver.OracleDriver");
```

2. 管理驱动程序类：DriverManager

DriverManager 类是 java.sql 包中用于数据库驱动程序管理的类，作用于用户和驱动程序之间。它跟踪可用的驱动程序，并在数据库和相应的驱动程序之间建立连接，也处理诸如驱动程序登录时间限制、登录和跟踪消息的显示等事务。一般的应用程序只使用它的 getConnection()方法，该方法用来建立与数据库的连接。

static Connection getConnection(String url, String username, String password)：通过指定的数据的 URL 及用户名、密码创建数据库连接。例如：

```
DriverManager.getConnection("jdbc:odbc:数据源名");
```

创建 SQL Server 数据库连接：

```
DriverManager.getConnection("jdbc:microsoft:sqlserver://localhost:1433;DatabaseNam
e=Test","sa","");
```

创建 Oracle 数据库连接：

```
DriverManager.getConnection("jdbc:oracle:thin:@10.10.11.153:1521:Test",user,password);
```

3. 数据库连接接口：Connection

Connection 是用来表示数据库连接的对象，提供 JDBC API 和 URL 指定的数据库系统之间的连接，也就是在它加载的 Driver 和数据库之间建立连接。对数据库的一切操作都是在这个连接基础上进行的。使用时必须创建一个 Connection Class 的实例，其中包括数据库信息。连接过程包括所执行的 SQL 语句和在该连接上所返回的结果。一个应用程序可以与单个数据库有一个或多个连接，或者可以与多个数据库有连接。

DriverManager 的 getConnection()方法，将建立在 JDBC URL 中定义的数据库的 Connection 连接上，代码如下：

```
Connection con=DriverManager. getConnection(url,login,password);
```

JDBC URL 提供了一种标识数据库的方法，JDBC URL 的标准语法由 3 部分组成，各部分间用冒号分隔：jdbc:<子协议>:<子名称>。JDBC URL 中的协议总是 jdbc，<子协议>为驱动程序名或数据库连接机制的名称，<子名称>为一种标识数据库的方法。

4. SQL 声明接口：Statement

Java 所有 SQL 语句都是通过陈述（Statement）对象实现的。Statement 用于在已经建立的连接的基础上向数据库发送 SQL 语句的对象。Statement 对象作为在给定 Connection 上执行 SQL 语句的容器，并且访问结果。Connection 接口提供了生成 Statement 的方法。

通过 Connection 对象的 createStatement 方法建立 Statement 对象：

```
Statement stmt=con.createStatement();
```

Statement 对象提供了 3 种执行 SQL 语句的方法。

ResultSet executeQuery(String sql)：执行 SELECT 语句，返回一个结果集。

int executeUpdate(String sql)：执行 update、insert、delete 等不需要返回结果集的 SQL 语句。它返回一个整数，表示执行 SQL 语句影响的数据行数。

boolean execute(String sql)：用于执行多个结果集、多个更新结果（或者两者都有）的 SQL 语句。它返回一个 boolean 值。如果第一个结果是 ResultSet 对象，则返回 true；如果是整数，就返回 false。

5. 查询结果接口：ResultSet

ResultSet 对象通过游标移动对给定 Statement 的结果进行访问，ResultSet 往往包含的是查询结果集，此接口抽象运行了 select 语句的结果，提供了逐行访问结果的方法，通过它来访问结果的不同字段。在 ResultSet 中隐含着一个指针，利用这个指针移动数据行，可以取得所要的数据，或对数据进行简单地操作。其主要的方法如表 5.7 所示。

表 5.7 ResultSet 方法

方 法	含 义
boolean absolute(int row)	将指针移动到结果集对象的某一行
void afterLast()	将指针移动到结果集对象的末尾
void beforeFrist()	将指针移动到结果集对象的头部
boolean first()	将指针移动到结果集对象的第一行
boolean next()	将指针移动到当前行的下一行
boolean previous()	将指针移动到当前行的前一行
boolean last()	将指针移动到当前行的最后一行

此外还可以使用一组 getXXX()方法，读取指定列的数据。XXX 是 JDBC 中 Java 语言的数据类型。这些方法的参数有两种格式，一是用 int 指定列的索引，二是用列的字段名（可能是别名）来指定列。例如：

```
String strName=rs.getString(2);
String strName=rs.getString("name");
```

5.3.2 编程模型及实例

在 Java 程序中要操作数据库，一般应该通过如下几步（利用 JDBC 访问数据库的编程步骤）。

（1）引用必要的包：

```
import java.sql.*;
```

（2）加载连接数据库的驱动程序类。

为实现与特定的数据库相连接，JDBC 必须加载相应的驱动程序类。通常可以采用 Class.forName()方法显式地加载一个驱动程序类，由驱动程序负责向 DriverManager 登记注册，并在与数据库相连接时，DriverManager 将使用此驱动程序。例如：

```
Class.forName("sun.jdbc.odbc.JdbcOdbcDriver");
```

（3）创建与数据源的连接。

```
String url="jdbc:odbc:数据源名";
Connection con=DriverManager.getConnection(url,"Login","Password");
```

注意：采用 DriverManager 类中的 getConnection()方法实现与 url 所指定的数据源建立连接，并返回一个 Connection 类的对象，以后对这个数据源的操作都是基于该 Connection 类对象。

（4）建立向数据库发送 SQL 语句对象。

在 JDBC 中查询数据库中的数据的执行方法可以分为 3 种类型，分别对应 Statement。PreparedStatement（预编译 SQL 语句）和 CallableStatement（主要用于执行存储过程）3 个接口。

（5）如果是查询，则取得 ResultSet 结果集中的数据。

（6）关闭查询语句及与数据库的连接（注意关闭的顺序，结果集对象、语句对象、连接对象）。

例 5-1 通过连接 MySQL 示例。

```
import java.sql.*;
public class Test
{
     public static void main(String[] args) throws SQLExpection
     {
          Connection con=null;
          try
          {
               Class.forName("com.mysql.jdbc.Driver");  //加载 mysql 驱动
               String url="jdbc:mysql://localhost:3306/test";
               con=DriverManager.getConnection(url,"sa","test"); //建立连接
          }
          catch (Exception ex)
          {
              ex.printStackTrace();
          }
          finally
          {
               try
               {
                    if(con!=null)
                         con.close();
               }
               catch(SQLException ignore)
               { }
          }
     }
}
```

当连接对象不在使用时，一定要关闭，如 con.close()。

例 5-2 运用前面介绍的知识能对图书管理系统中的借阅进行管理，步骤如下所述。

（1）建立数据库操作类：DataBaseManager.java。

```
import java.sql.*;
public class DataBaseManager
{
    Connection con;
    ResultSet rs;
    Statement stmt;
    public DataBaseManager()
```

```
{
    try
    {
        Class.forName( "sun.jdbc.odbc.JdbcOdbcDriver" ); //加载 Oracle 驱动
        con=DriverManager.getConnection("jdbc:oracle:thin:@test:1521:myoracle",
                                         "test","test"); //建立连接
        stmt=con.createStatement();         //创建语句
    }
    catch ( ClassNotFoundException cnfex ) {
        System.err.println("Failed to load JDBC/ODBC driver." );
        cnfex.printStackTrace();
        System.exit( 1 );                   // terminate program
    }
    catch(SQLException sqle)
    {
        System.out.println(sqle.toString());
    }
}
public ResultSet getResult(String strSQL)
{
    try{
        rs=stmt.executeQuery(strSQL);       //执行查询操作
        return rs;
    }
    catch(SQLException sqle)
    {
        System.out.println(sqle.toString());
        return null;
    }
}

public boolean updateSql(String strSQL)
{
    try{
        stmt.executeUpdate(strSQL);         //执行更新、删除、修改操作
        con.commit();
       return true;
    }
    catch(SQLException sqle)
    {
        System.out.println(sqle.toString());
        return false;
    }
}

public void closeConnection()
{
    try
    {
        rs.close();
        stmt.close();
        con.close();
    }
```

```
        catch(SQLException sqle)
        {
            System.out.println(sqle.toString());
        }
    }
}
```

（2）执行借阅操作类。

```
import java.sql.*;
public class BorrowBook
{
    DataBaseManager db=new DataBaseManager();
    void borrow(String sno ,String bno ,String bdate , String rdate )
    {
        try
        {
         String strSQL="insert into bookBrowse" +
         "(studentname,bookname,borrowdate,returndate)" +
        "values('"+ sno + "','" + bno + "','" + bdate + "','" + rdate +"')";
            //表名为 bookBrowse ，表结构：学生名、书名、借阅日期、归还日期
            if(db.updateSql(strSQL)) //执行插入操作
            {
                System.out.println("借阅完成！");
                db.closeConnection();
            }
         }
         catch(Exception ex)
         {
            System.out.println(ex.toString());
         }
    }
}
```

（3）执行查询图书操作类。

```
    try
      {
            String strSQL="select studentname,bookname,borrowdate from"+
                            " bookBrowse where studentName='"+ sno +"'" ;
            rs=db.getResult(strSQL); //执行查询
            if(!rs.first()){
                    System.out.println("此学生没有借过书!");
                    db.closeConnection();
                    return false;
            }else{
                    while(rs.next()) //循环取得结果集中记录
                    {
                      System.out.println("学生姓名："+rs.getString("studentname "));
                      System.out.println("图书名："+rs.getString("bookname "));
                      System.out.println("借阅日期："+rs.getString("borrowdate "));
                    }
                    db.closeConnection();
                    return true;
                }
      }
```

```
            catch(Exception ex)
            {
                System.out.println(ex.toString());
                return false;
            }
        }
    }
```

（4）执行删除图书操作类。

```
import java.sql.*;
public class DeleteBook
{
    DataBaseManager db=new DataBaseManager();
    void Delete (String sno )
    {
        try
        {
            String strSQL="delete from bookBrowse where studentname='"+ sno +"'";
            if(db.updateSql(strSQL))  //执行删除操作
            {
                System.out.println("删除完成! ");
                db.closeConnection();
            }
        }
        catch(Exception ex)
        {
            System.out.println(ex.toString());
        }
    }
}
```

（5）执行修改图书操作类。

```
import java.sql.*;
public class UpdateBook
{
    DataBaseManager db=new DataBaseManager();
    void Update (String sno ,String bno)
    {
        try
        {
            String strSQL="update bookBrowse set bno ='" + bno + "' where "+
                          " studentname='"+ sno +"'" ;
            if(db.updateSql(strSQL))  //执行修改操作
            {
                System.out.println("修改完成! ");
                db.closeConnection();
            }
        }
        catch(Exception ex)
        {
                System.out.println(ex.toString());
        }
    }
}
```

5.3.3 解决中文乱码问题

这里讨论的数据库乱码问题主要可分为从 jsp 页面添加中文数据到数据库的乱码和取出数据是乱码的问题。

首先讨论数据保存至 MySQL 数据库中是乱码的问题。

（1）获得 MySQL 的字符集。查看字符集的命令是：

```
Mysql>show variables like “character_set_%”;
```

（2）在 JSP 页面制定编码格式，保证与 mysql 的字符集一致，例如，在 jsp 页面声明：

```
<%@ page language="java" import="java.util.* " pageEncoding="GB2312"%>
```

如果添加数据的页面和显示数据的页面编码格式不一致，则会出现添加到数据库中无乱码，但从数据库中显示则有乱码的情况。出现这种情况时，可以在设定连接数据库的 URL 时指定字符编码。例如：

```
String url="jdbc:mysql://localhost:3306/test";
useUnicode=true&characterEncoding="GB2312";
```

采用这种方式得到的数据，使用的字符编码即为指定的 GB2312。对于不同的数据库，URL 中字符编码的设置形式会有不同的情况。如果没有用上述方法在 URL 中设定字符编码，则应该首先知道 JDBC 驱动程序中使用的默认字符编码（一般为“ISO-8859-1”），然后对从数据库中取出的数据进行转换，例如在显示页面的 jsp 中对得到数据做如下转换：

```
String result = new String(type.getBytes("iso-8859-1"), "gb2312");
```

最后需要特别注意的是，不同数据库的驱动程序，或者同一数据库的不同版本的驱动程序的默认编码字符都可能存在不一样的情况。

思考与习题

1. 填空题：JDBC（Java Database Connectivity）是________的 API（application PRogramming Interface），与 ODBC 一样，JDBC 是建立在 X/Open SQL CLI 基础上的。
2. 填空题：数据库系统编程模型中的两层模型，是指________架构，即应用层和数据库层。
3. 填空题：JDBC 有 3 种产品组件：________、________和________。
4. 简答题：简述关系型数据库和非关系型数据库的区别。
5. 简答题：简述 JDBC 访问数据库的编程步骤。
6. 编程题：下载安装 MySQL，并练习数据库查询语言。
7. 编程题：查询某个学生某门课程的成绩。
8. 编程题：插入学生的选课情况，并根据学号将选课结构查询显示。

第 6 章 JSP、Servlet 和 JavaBean

JSP+Servlet+JavaBean 开发模式是常见的开发模式，其遵循 MVC 的设计理念。其中 JSP 作为视图层为用户提供与程序交互的界面，JavaBean 作为模型层封装实体对象及业务逻辑，Servlet 作为控制层接收各种业务请求，并调用 JavaBean 模型组件对业务逻辑进行处理。

本章将简要介绍 JSP、Servlet，以及 JavaBean 技术。

6.1 JSP 基础

6.1.1 JSP 技术概述

在 Internet 发展的最初阶段，HTML 只能在浏览器中展现静态的文本或图像信息，无法满足人们对信息丰富性和多样性的强烈需求。随着 Internet 和 Web 技术应用到商业领域，Web 技术功能越来越强大。目前，解决 Web 动态网站的开发技术有很多，如 Servlet、JSP、ASP、PHP 等，都得到了广泛的应用，JSP，是其中的佼佼者。

JSP（Java Server Pages）是由 Sun Microsystem 公司于 1999 年推出的新技术，是基于 Java Servlet 及整个 Java 体系的 Web 开发技术。利用这一技术，可以建立先进的安全的和跨平台的动态网站。在传统的网页 HTML 文件（*.htm，*.html）中加入 Java 程序片段（Scriptlet）和 JSP 标记，就构成了 JSP 网页（*.jsp）。Web 服务器在收到访问 JSP 网页的请求时，首先执行其中的程序片段，然后将执行结果以 HTML 格式返回给客户。程序片段可以操作数据库、重新定向网页、发送 E-mail 等，这就是建立动态网站所需要的功能。JSP 所有程序操作都在服务器端执行，网络上传送给客户端的仅是得到的结果，对客户浏览器的要求很低。

在 JSP 页面运行时，需要有相应的编译器编译和解释器来执行这些 Java 代码。执行 Java 代码，需要下载和安装 Java 的 JDK 开发工具包。而开发 JSP 需要的软件主要有两大类：Web 应用服务器与数据库。JSP 技术可以使用的服务器有很多，如 Tomcat、Jboss、Resin、WebLogic 等。每个服务器都有自己的特点，其应用方面也不相同。其中 Tomcat 服务器在中、小型的 JSP 网站上应用比较广泛，具有和 JSP 技术结合紧密等特点。Tomcat 的下载与安装请参考本书 1.2.5 小节中的详细介绍。

1. JSP 的主要优势

（1）数据内容和显示分离：使用 JSP 技术，Web 开发人员可以使用 HTML 或者 XML 标记来设计和格式化最终页面。使用 JSP 标记或者小脚本来产生页面上的动态内容，产生内容的逻辑被封装在标记和 JavaBeans 群组件中，并且捆绑在小脚本中，所有的脚本在服务器端执行。如果核心逻辑被封装在标记和 Beans 中，那么其他人，如 Web 管理人员和页面设计者，能够编辑和使用 JSP 页面，而不影响内容的产生。在服务器端，JSP 引擎解释 JSP 标记，产生所请求的内容（例如，通过存取 JavaBeans 群组件，使用 JDBC 技术存取数据库），并且将结果以 HTML（或者 XML）页面的形式发送回浏览器。这有助于作者保护自己的代码，而且又保证任何基于 HTML 的 Web 浏览器的完全可用性。

（2）可重用组件：绝大多数 JSP 页面依赖于可重用且跨平台的组件（如 JavaBeans 或者 Enterprise JavaBeans）来执行应用程序所要求的更为复杂的处理。开发人员能够共享和交换执行普通操作的组件，或者使得这些组件为更多的使用者或者用户团体所使用。基于组件的方法加速了总体开发过程，并且使各种群组织在他们现有的技能和开发努力的优化结果中得到平衡。

（3）采用标记简化页面开发：JSP 技术封装了许多功能，这些功能是在易用的、与 JSP 相关的 XML 标记中生成动态内容所需要的。标准的 JSP 标记能够存取和实例化 JavaBeans 组件，设定或者检索群组件属性，下载 Applet，以及执行用其他方法更难于编码和耗时的功能。

2. JSP 开发模式

JSP 网站开发技术，经常使用下面几种组合开发网站，包括纯粹 JSP 技术实现、JSP+JavaBean 实现、JSP+JavaBean+Servlet 实现、J2EE 实现等。不同的开发组合，可以称为不同的设计模式，最常用的技术是 JSP+JavaBean+Servlet。该技术的组合很好地实现了 MVC（Model-View-Controller）模式。MVC 应用程序由模式-视图-控制器这 3 个部分组成，其实现将在下一章具体介绍。在该开发模式中，JSP 主要用来处理显示模块，即 View；JavaBean 用来处理业务逻辑，即 Model；Servlet 作为控制器，即 Controller。MVC 模式最早是 Smalltalk 语言研究团提出的，应用于用户交互应用程序。

6.1.2 JSP 基本语法

JSP 的基本语法包括 JSP 元素和 JSP 指令。

1. JSP 元素

JSP 文件中有 5 类元素，分别是模板元素、注释、脚本元素、指令元素和动作元素。

（1）模板元素：模板元素是指 JSP 的静态 HTML 内容。它对 JSP 的显示是非常必要的，但对编程人员来说则很少关注，因为这部分任务主要由网页美工来完成。

可以说模板元素是网页的框架，直接影响着网页的结构和美观程度。在 JSP 编译时，直接将这些模板元素编译到 Servlet 里。当客户请求此 JSP 时，会把这些模板元素原样发送到客户端。

（2）注释：JSP 中的注释有多种情况，有 JSP 自带的注释规范，也有 HTML/XML 的注释规范。

HTML/XML 注释：用于在客户端显示一个注释。语法结构为：

```
<!- -comment[<%=expression %>]- ->
```

这种注释和 HTML 很像，唯一不同的是 HTML/XML 注释中可以使用表达式。

隐藏注释：写在 JSP 程序中但不发送给客户。语法结构为：

```
<%- -comment- -%>
```

用隐藏注释标记的字符在 JSP 的编译时会被忽略，也就是说 JSP 编译器在编译时不会对<%--和--%>之间的语句进行编译。

Scriptlets 中的注释：Java 中的注释在 Scriptlets 中同样适用。

//：以“//”开头换行符结尾，表示单行注释。

/*、*/：以“/*”开头“*/”结尾，表示单行或多行注释。

/**、*/：以“/**”开头“*/”结尾，表示文档注释。

（3）脚本元素：脚本元素主要包括声明、表达式和 Scriptlets。它将 JSP 的所有一切都集中在一起，通常是用 Java 写的。

（a）声明：在 JSP 中声明是一段 Java 代码，它用来定义在产生的类文件中的类的属性和方法。然后在 JSP 中可以任意使用。格式：<%!....%>，当然也可以把这段程序代码单独做成一个小 JSP，用<%@ include file=""%>包含加入页面。

（b）表达式：在 JSP 请求处理阶段计算它的值，把获得的结果转换成字符串，并与模板数据组合在一起，位置就在它所处的 JSP 中的位置。语法结构为：

```
<%=expression%>
```

（c）Scriptlets：一段可以在处理请求时间执行的 Java 代码，可以产生输出的结果。

（4）指令元素：用于从 JSP 发送一个信息到容器中，用来设置全局变量、声明类和输出内容的类型等。指令元素并不向客户产生任何输出，所有的指令在 JSP 整个文件中有效。JSP 中指令元素有 3 种，分别是页面（page）指令、include 指令和 taglib 指令。指令格式：

```
<% @ directivename attribute="value",attribute="value"%>
```

（5）动作元素：不同于指令元素，动作元素是在请求处理阶段起作用，且采用了 XML 语法。容器在处理 JSP 时，根据动作元素的标记来进行特殊处理。格式：

```
<prefix:tag attribute=value attribute-list.../>
```

2. JSP 指令

JSP 包含 3 个编译指令和 7 个动作指令。3 个编译指令为：page、include、taglib；7 个动作指令为：jsp:forward、jsp:param、jsp:include、jsp:plugin、jsp:useBean、jsp:setProperty、jsp:getProperty。JSP 指令容器的消息，使程序员能够指定页面的设置、包含其他资源中的内容和指定 JSP 中使用的定制标记库。它是那些发送给 JSP 影响 Servlet 类的整体结构，语法如下：

```
<%@  directive attribute="value"  %>
```

另外，也可以把同一指令的多个属性结合起来，例如：

```
<%@ directive attribute1="value1" attribute2="value2" ... attributeN="valueN" %
```

（1）编译指令

（a）page 指令：主要功能是设定整个 JSP 网页的属性和相关功能，page 指令的语法格式如下：

```
<%@  page
[language="java"]
[extends="package.class"]
[import="{package.class/package.*},…"]
[session="true/false"]
[buffer="none/8kb/sizekb"]
[autoFlush="true/false"]
[info="text"]
[errorPage="relativeURL"]
[contentType="mimeType[;charset=characterSet]"  /"text/html;charset=ISO-8859-1" ]
[isErrorPage="true/false"]
%>
```

page 指令用来定义下面一个或多个属性，这些属性大小写敏感。

language 属性：一般情况就是 Java，代表 JSP 页面使用的脚本语言。

extends 属性：确定 JSP 程序编译时所产生的 Java 类，需要继承的父类，或者需要实现的接口的全限定类名。

import 属性：用来导入包，下面几个包是默认自动导入的，不需要显式导入。默认导入的包有：

```
java.lang.*
javax.servlet.*
javax.servlet.jsp.*
javax.servlet.http.*
```

session 属性：设定这个 JSP 页面是否需要 HTTP Session。

buffer 属性：指定输出缓冲区的大小。输出缓冲区的 JSP 内部对象：out 用于缓存 JSP 页面对客户浏览器的输出，默认值为 8kb，可以设置为 none，也可以设置为其他值，单位为 kb。

autoFlush 属性：当输出缓冲区即将溢出时，是否需要强制输出缓冲区的内容。设置为 true 时为正常输出；如果设置为 false，会在 buffer 溢出时产生一个异常。

info 属性：设置该 JSP 程序的信息，也可以看作其说明，可以通过 Servlet. getServletInfo()方法获取该值。如果在 JSP 页面中，可直接调用 getServletInfo()方法获取该值，因为 JSP 页面的实质就是 Servlet。

errorPage 属性：指定错误处理页面。

isErrorPage 属性：设置 JSP 页面是否为错误处理程序。如果该页面本身已是错误处理页面，则无须使用 errorPage 属性。

contentType 属性：用于设定生成网页的文件格式和编码方式，即 MIME 类型和页面字符集类型，默认的 MIME 类型是 text/html；默认的字符集为 ISO-8859-1。

（b）include 指令：表示在 JSP 编译时插入一个包含文本或代码的文件，这个包含的过程是静态的（即编译时完成）。其包含的文件可以是 JSP 网页、HTML 网页、文本文件，或是一段 Java 程序文本。include 指令语法格式如下：

```
<%@ include file="relative URL %>
```

include 指令只有一个属性 file，relative URL 表示指定 file 的路径。这里所指定的 URL 是和发出引用指令的 JSP 页面相对的 URL，然而，与通常意义上的相对 URL 一样，可以利用以“/”开始的 URL 告诉系统把 URL 视为从 Web 服务器根目录开始。包含文件的内容也是 JSP 代码，即包含文件可以包含静态 HTML、脚本元素、JSP 指令和动作。

（c）taglib 指令：定义一个标签库及其自定义标签的前缀。taglib 指令的语法格式如下：

```
<%@ taglib uri="URIToTagLibrary" prefix="tagPrefix" %>
```

（2）动作指令

JSP 动作利用 XML 语法格式的标记来控制 Servlet 引擎的行为。利用 JSP 动作可以动态地插入文件、重用 JavaBean 组件、把用户重定向到另外的页面、为 Java 插件生成 HTML 代码。

JSP 动作元素有两种格式：

```
<prefix:tag attribute=value attribute-list…/>或者
<prefix:tag attribute=value attribute-list…>
…
</ prefix:tag >
```

（a）forward 指令：用于将页面响应控制转发给另外的页面。既可以转发给静态的 HTML 页面，也可以转发到动态的 JSP 页面，或者转发到容器中的 Servlet。forward 指令的格式如下：

```
<jsp:forward page="{relativeURL |<%=expression%>}">
{<jsp:param…/>}
</jsp:forward>
```

（b）param 指令：param 指令用于设置参数值，这个指令本身不能单独使用，它可以和以下几个指令合起来使用：jsp:include、jsp:forward 和 jsp:plugin。

（c）include 指令：它是一个动态的指令，可以用于导入某个页面。它的导入会每次检查被导入页面的改变。include 指令的使用格式：

```
<jsp:include page="{relativeURL |<%=expression%>}" flush="true">
<jsp:param name="paramName" value="paramValue"/>
</jsp:include>
```

（d）plugin 指令：plugin 指令主要用于下载服务器端的 JavaBean 或 Applet 到客户端执行。由于程序在客户端执行，因此客户端必须安装虚拟机。

（e）useBean、setProperty 和 getProperty 指令：这 3 个指令都是与 JavaBean 相关的指令，其中 useBean 用于在 JSP 页面初始化一个 Java 实例，setProperty 用于修改 JavaBean 实例的属性，getProperty 用于获取 JavaBean 实例的属性。

useBean 的语法格式：

```
<jsp:useBean id="" class="" scope="page | request |session | application" >
```

setProperty 的语法格式：

```
<jsp:setProperty name="" property="" value="" />
```

getProperty 与 setProperty 的语法格式类似。

6.1.3 JSP 内置对象

1. page 对象

page 对象代表 JSP 本身，对象类型为 java.lang.Object，它代表 JSP 被转译后的 Servlet，因此，它可以调用 Servlet 类所定义的方法，作用域为页面的执行期。不过实际上，page 对象很少在 JSP 中使用。

例 6-1 page 对象在 JSP 中的使用。

```
<%@ page info="JSP内置page对象" contentType="text/html; charset=GB2312" %>
<html>
<head>
<title> PageDemo.jsp</title>
</head>
<body>
Page Info = <%=((javax.servlet.jsp.HttpJSPPage)page).getServletInfo()%>
</body>
</html>
```

在浏览器中执行的结果为显示 Page Info=JSP 内置 page 对象。

2. config 对象

config 对象里存放着一些 Servlet 初始的数据结构，config 对象和 page 对象一样都很少被用到。config 对象实现于 javax.servlet.ServletConfig 接口，作用域为页面的执行期。它共有下列 4 种方法：

```
public String getInitParameter(name)   //config对象取得 Servlet 初始参数值
public java.util.Enumeration getInitParameterNames( )   // config 对象取得 Servlet
```

初始参数值

```
public  ServletContext  getServletContext( )  //config 对象取得 Servlet 上下文信息
public  Sring  getServletName( )   //config 对象取得 Servlet 名
```

例 6-2

在 web.xml 中有如下的信息：

```
<init-param>
    <param-name>ParaTest</param-name>
    <param-value>
        10
    </param-value>
</init-param>
```

在 JSP 页面中可以使用：

```
<% String  propertyFile = (String)config.getInitParameter("propertyFile"); %>
```

取得名称为 ParaTest，其值为 10 的参数。

3. request 对象

request 对象包含所有请求的信息，如请求的来源、标头、cookies 和请求相关的参数值等，类型为 javax.servlet.ServletRequest 子类，作用域为用户请求期。

例 6-3　request 对象在 JSP 中的应用。

```
<html>
<head>
    <title>Test.html</title>
    <meta  http-equiv="Content-Type"  content="text/html;  charset=GB2312">
</head>
<body>
<form  action="Result.jsp"  method="GET">
    Name: <input  type="text"  name="姓名"  size="20"  maxlength="20"><br>
    <input  type="submit"  value="提交">
</form>
</body>
</html>
```

在 Result.jsp 页面中包含“<%=　request.getParameter("姓名")　%>”语句，则会显示输入的值。

4. response 对象

response 对象主要将 JSP 处理数据后的结果传到客户端，是 JSP 页面的响应，类型为 javax.servlet.ServletResponse 子类，作用域为用户请求期。

例如，使用“response.setIntHeader("Refresh", 180)”语句，就是告诉浏览器，每隔 3min 重新加载此网页；使用“response.setHeader("Refresh", " 10; URL=http://Test/Mypath")”语句，表示过 10s 后，调用浏览器转到“http://Test/Mypath”的网页。

5. out 对象

out 对象能把结果输出到网页上，代表输出流对象，作用域为页面执行期。最常使用 out.println(String name)和 out.print(String name)，它们两者最大的差别在于 println()在输出的数据后面会自动加上换行的符号。

例 6-4　out 对象在 JSP 中的应用。

```
<%@  page  info="JSP 内置 out 对象"  contentType="text/html;charset=GB2312"  %>
```

```
<html>
<head>
<title> OutDemo.jsp</title>
</head>
<body>
<%= out.getBufferSize(); %>
</body>
</html>
```

在浏览器中执行的结果为显示开始缓冲区的默认大小。

6. session 对象

session 对象表示目前用户的会话（session）状况，储存或取得用户相关的数据，session 对象实现 javax.servlet.http.HttpSession 接口，作用域为会话期。

例 6-5　session 对象在 JSP 中的应用。

```
<%@ page contentType="text/html;charset=GB2312" %>
<html>
<head>
    <title>Login.jsp</title>
</head>
<body>
<form action="Result.jsp" method="GET">
    Name: <input type="text" name="姓名" size="20" maxlength="20"><br>
    <input type="submit" value="提交">
</form>
<%  if (request.getParameter("姓名") != null)
    {
        String Name = request.getParameter("姓名");
        if (Name.equals("test"))
              session.setAttribute("Para", "OK");
        }
    %>
    </body>
</html>
```

在 Login.jsp 的程序中，要求用户分别输入姓名，如果输入的名称为 test 时，就把名称为 Para、其值为 OK 的属性加入到 session 对象当中。

7. application 对象

application 对象实现 javax.servlet.ServletContext 接口，主要用于取得或更改 Servlet 的设定，作用域在整个 Web 应用程序运行期。

application 对象最常被用来存取环境的信息，因为环境的信息通常都储存在 ServletContext 中，所以常利用 application 对象来存取 ServletContext 中的信息。

例如，<%=application.getServerInfo() %>。

8. pageContext 对象

pageContext 对象能够存取其他隐含对象，类型为 javax.servlet.jsp.PageContext，作用域为 Page 页面执行期。

例 6-6　pageContext 对象在 JSP 中的应用。

```
<%@ page info="内置 pageContext 对象" contentType="text/html;charset=GB2312" %>
<html>
```

```
<head>
<title> OutDemo.jsp</title>
</head>
<body>
<%
    Enumeration  e =
        pageContext.getAttributeNamesInScope(PageContext.APPLICATION_SCOPE  );
    while  (e.hasMoreElements())
     out.println(e.nextElement())
%>
</body>
</html>
```

pageContext.getAttributeNamesInScope()会传递所有指定范围的属性名称，利用 Enumeration 对象 e 来收集所有属性范围为 Application 的数据，再一一地取出打印出来。

9. exception 对象

当 JSP 网页产生异常时交由 exception 对象做处理。类型为 java.lang.Throwable，作用域为页面执行期。

例 6-7　exception 对象在 JSP 中的应用。

```
<%@  page  contentType="text/html;charset=GB2312"  isErrorPage="true"  %>
<html>
<head>
<title> OutDemo.jsp</title>
</head>
<body>
<%=  exception  %>
<%=  exception.getMessage()  %>
<%=  exception.getLocalizedMessage()  %>
</body>
</html>
```

6.2　Servlet 基础

6.2.1　Servlet 简介

Servlet 是使用 Java 语言编写的服务器端程序，它能够接受客户端的请求并产生响应。最早支持 Servlet 技术的是 JavaSoft 的 Java Web Server。Servlet 发展至今，已经发展为一门成熟的技术，在 Java Web 的应用编程中，掌握 Servlet 是非常重要的。

Java Servlet 通常情况下与使用 CGI（Common Gateway Interface，公共网关接口）实现的程序可以达到异曲同工的效果。但是相比于 CGI，Servlet 有以下几点优势。

- 性能明显更好。
- Servlet 在 Web 服务器的地址空间内执行。这样它就没有必要再创建一个单独的进程来处理每个客户端请求。
- Servlet 是独立于平台的。
- 服务器上的 Java 安全管理器执行了一系列限制，以保护服务器计算机上的资源。因此，

Servlet 是可信的。

- Java 类库的全部功能对 Servlet 来说都是可用的。它可以通过 sockets 和 RMI 机制与 applets、数据库或其他软件进行交互。

1. Servlet 架构

图 6.1 显示了 Servlet 在 Web 应用程序中的架构。

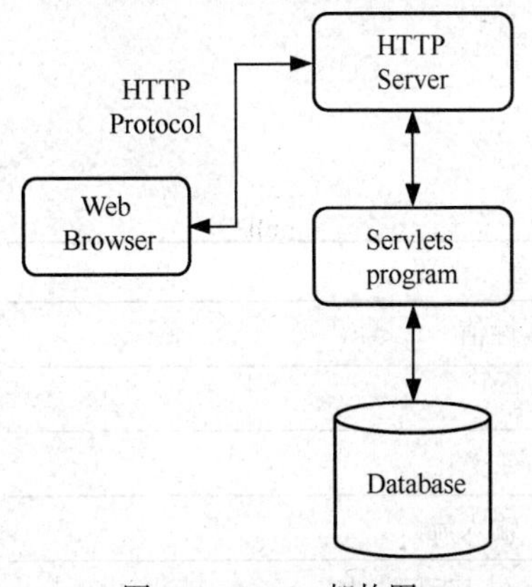

图 6.1 Servlet 架构图

2. Servlet 任务

Servlet 的运行需要容器的支持，它通常被部署在容器中，其中 Tomcat 是最常用的 JSP/Servlet 容器。Servlet 容器负责 Servlet 和客户的通信及调用 Servlet 的方法，其通信采用"请求/响应"模式。

Servlet 的主要功能在于交互式地浏览和修改数据，生成动态的 Web 内容。这个过程为：

（1）客户端发送请求至服务器端；

（2）服务器将请求信息发送至 Servlet；

（3）Servlet 生成响应内容并将其传给服务器。响应内容动态生成，通常取决于客户端的请求；

（4）服务器将响应返回给客户端。

6.2.2 Servlet 的类与接口

Servlet 的框架是由两个包组成的：javax.servlet 和 javax.servlet.http。前者中定义了所有的 servlet 类都必须实现或扩展的通用接口和类；而后者中定义了采用 HTTP 协议通信的 HttpServlet 类。

Servlet 框架的核心是 javax.servlet.Servlet 接口，所有的 Servlet 都必须实现这一接口。Servlet 接口中定义了 5 个方法：init 方法用于初始化 Servlet，destroy 方法用于销毁 Servlet，getServletInfo 方法用于获取信息，getServletConfig 方法用于获得 Servlet 配置相关信息，service 方法负责相应客户的请求。其中，init 方法、service 方法和 destroy 方法代表了 Servlet 的生命周期。

1. GenericServlet 抽象类

为了简化 servlet 的编写，在 javax.servlet 包中提供了一个抽象类 GenericServlet，它给出了除 service()方法以外的简单实现。

GenericServlet 类中的 service 方法是抽象方法，该方法的声明形式如下：

```
public abstract void service(ServletRequest request, ServletResponse response)
throws ServletException, IOException;
```

2. HttpServlet 抽象类

HttpServlet 主要是应用于 HTTP 协议的请求和响应，为了快速开发 HTTP 协议的 servlet，sun 提供了一个继承自 GenericServlet 的抽象类 HttpServlet，用于创建适合 Web 站点的 HTTP Servlet。该类中实现了 Servlet 接口的 service 方法，声明如下：

```
public  void service(HttpServletRequest request, HttpServletResponse response)
throws ServletException, IOException;
```

在 HttpServlet 的 service 方法中，doGet 方法用于处理和响应 Get 请求，doPost 方法用于处理和响应 Post 请求。首先从 HttpServletRequest 对象中获取 HTTP 的请求方式，再调用相应的方法。

3. ServletRequest 接口

ServletRequest 接口封装了客户端请求的细节。它与协议无关，并有一个指定 HTTP 的子接口。

ServletRequest 主要做以下处理：

（1）找到客户端的主机名和 IP 地址；

（2）检索请求参数；

（3）取得和设置属性；

（4）取得输入和输出流；

表 6.1 给出了 ServletRequest 接口的几种方法。

表 6.1　ServletRequest 接口的方法

方　法	意　义
getAttribute	根据参数给定的属性名返回属性值
getContentType	返回客户请求数据 MIME 类型
getCharacterEncoding	返回请求所用的字符编码
getInputStream	返回与请求相关的（二进制）输入流
getParameter	返回指定输入参数，如果不存在，返回 null
getServerName	返回处理请求的服务器的主机名
getServerPort	返回接收主机正在侦听的端口号
getRemoteAddr	返回客户端主机的数字型 IP 地址
getRemoteHost	返回客户端主机名

4. ServletResponse 接口

ServletResponse 对象将一个 servlet 生成的结果传到发出请求的客户端。ServletResponse 操作主要是作为输出流及其内容类型和长度的包容器，它由 servlet 引擎创建。

表 6.2 给出 ServletResponse 接口的几种方法。

表 6.2　ServletResponse 接口的方法

方　法	意　义
getCharacterEncoding	返回响应使用字符解码的名字
getOutputStream	返回用于将返回的二进制输出写入客户端的流
getWriter	返回可以向客户端发送字符数据的 PrintWrite 对象
setContentLength	设置内容体的长度
setContentType	设置内容类型

6.2.3 Servlet 生命周期

由于 Servlet 部署在容器里，其生命周期由容器管理。Servlet 的生命周期可以分为 4 个阶段：创建对象阶段、初始化阶段、响应客户请求阶段和终止阶段。

（1）创建对象阶段。当 Web 客户请求 Servlet 服务时，容器环境加载一个 Java Servlet 类，并创建一个或多个 Servlet 对象实例，并将这些对象实例加入 Servlet 实例池中。

（2）初始化阶段。环境容器调用 Servlet 的初始化方法 init()进行初始化。首先给 init()方法传入一个 ServletConfig 对象，该对象中包含了初始化参数和容器环境的信息，并负责向 Servlet 传递数据。

在整个 Servlet 生命周期中，init()方法只会被调用一次。它在第一次创建 Servlet 时被调用，在后续每次用户请求时不再调用。因此，它是用于一次性初始化，就像 Applet 的 init 方法一样。init 方法的定义如下：

```
public void init() throws ServletException {
  // 初始化代码...
}
```

（3）响应客户请求阶段。针对不同的客户请求，Servlet 容器创建特定请求的 ServletRequest 对象和 ServletResponse 对象，并将其传递给 service 方法。ServletRequest 对象将客户请求信息交由 service 处理，ServletResponse 对象向客户返回结果。service 方法可以被多次调用，各调用过程运行在不同的线程中且互不干扰。service()方法检查 HTTP 请求类型（GET、POST、PUT、DELETE 等），并在适当的时候调用 doGet、doPost、doPut，doDelete 等方法。service 方法的定义如下：

```
public void service(ServletRequest request, ServletResponse response)
    throws ServletException, IOException{
}
```

（4）终止阶段。当服务器和容器关闭时，会调用 destroy()方法释放 Servlet 占用的资源，并进行关闭前的处理。destroy()方法只会被调用一次，destroy()方法可以关闭 Servlet 的数据库连接、停止后台线程、把 Cookie 列表或点击计数器写入到磁盘，并执行其他类似的清理活动。

在调用 destroy()方法之后，servlet 对象被标记为垃圾回收。destroy 方法定义如下所示：

```
public void destroy() {
  // 终止化代码...
}
```

6.2.4　Servlet 表单数据

当浏览器需要传递数据到 Web 服务器时，可以使用两种方法将这些信息传递，分别为 Get 方法和 Post 方法。

（1）Get 方法：向页面请求发送已编码的用户信息。页面和已编码的信息中间用?字符分隔，如下所示：

```
http://www.test.com/hello?key1=value1&key2=value2
```

Get 方法是默认地从浏览器向 Web 服务器传递信息的方法，它会在浏览器的地址栏中产生一个很长的字符串。Get 方法有大小限制：请求字符串中最多只能有 1024 个字符。这些信息使用 QUERY_STRING 头传递，并可以通过 QUERY_STRING 环境变量访问，Servlet 使用 doGet()方法处理这种类型的请求。

（2）Post 方法：Post 方法打包信息的方式与 Get 方法基本相同，但是 Post 方法不是把信息作为 URL 中?字符后的文本字符串进行发送，而是把这些信息作为一个单独的消息，消息以标准输出的形式传到后台程序。Servlet 使用 doPost()方法处理这种类型的请求。

Servlet 处理表单数据，这些数据会根据不同的情况使用不同的方法自动解析。

- getParameter()：request.getParameter()方法可以获取表单参数的值。
- getParameterValues()：若参数出现一次以上，则调用该方法，并返回多个值，例如复选框。
- getParameterNames()：若想要得到当前请求中的所有参数的完整列表，则调用该方法。

1. 使用 URL 的 Get 方法

如下所示的 URL，将使用 Get 方法向程序传递参数 name 的值：

```
http://localhost:8080/HelloForm?name=mm
```

在 Servlet 程序中，可以使用 getParameter()方法，可以很容易地访问传递的信息，如下代码片段所示：

```
out.println(request.getParameter("name") )
```

当完整的程序编译完成并启动 Tomcat 服务器后，在浏览器的地址中输入“http://localhost:8080/HelloForm?name=mm”，将会得到参数 name 的值 mm。

2. 使用表单的 Get 方法

除了在 URL 中传递参数至后台程序，也可以在前台表单中输入参数，传递至后台程序，下面是一个使用 HTML 表单和提交按钮传递参数的实例。

例 6-8 HTML 表单和提交按钮传递参数的实例。

```
<html>
<body>
<form action="XXX" method="GET">
名字: <input type="text" name="first_name">
<br />
姓氏: <input type="text" name="last_name" />
<input type="submit" value="提交" />
</form>
</body>
</html>
```

将该 HTML 文件命名为 hello.htm，并在“<Tomcat-installation-directory>/webapps/ROOT”目录下放置该文件。运行“http://localhost:8080/hello.htm”时，则会出现表单。自行输入姓氏和名字单击“提交”按钮，在 Servlet 程序中仍使用 getParameter()方法，获取到提交的姓氏和名字两个参数。注意在 Servlet 程序中，起关键性作用的 getParameter()方法，要写在 Get 方法中，Get 方法结构如下：

```
public void doGet(HttpServletRequest request,
                  HttpServletResponse response)
        throws ServletException, IOException {
}
```

3. 使用表单的 Post 方法

下面给出一个完整的 Servlet 程序，它既可以使用 Get 方法，也可以 Post 方法来处理由 Web 浏览器给出的输入。

例 6-9 Get、Post 方法在 Servlet 中的应用。

```
import java.io.*;
import javax.servlet.*;
import javax.servlet.http.*;          // 导入必需的 Java 库
public class HelloForm extends HttpServlet {    // 扩展 HttpServlet 类
public void doGet(HttpServletRequest request,    // 处理 Get 方法请求的方法
HttpServletResponse response)
    throws ServletException, IOException
  {
```

```
        response.setContentType("text/html");
           PrintWriter out = response.getWriter();
         String title = "Using GET Method to Read Form Data";
        String docType =
        "<!doctype html public \"-//w3c//dtd html 4.0 " +
        "transitional//en\">\n";
        out.println(docType +
                "<html>\n" +
                "<head><title>" + title + "</title></head>\n" +
                "<body bgcolor=\"#f0f0f0\">\n" +
                "<h1 align=\"center\">" + title + "</h1>\n" +
                "<ul>\n" +
                "  <li><b>名字</b>: "
                + request.getParameter("first_name") + "\n" +
                "  <li><b>姓氏</b>: "
                + request.getParameter("last_name") + "\n" +
                "</ul>\n" +
                "</body></html>");
    }
    public void doPost(HttpServletRequest request,       // 处理 Post 方法请求的方法
        HttpServletResponse response)
        throws ServletException, IOException {
       doGet(request, response);
    }
}
```

编译部署上述的 Servlet，并使用带有 Post 方法的 hello.htm 进行测试，如下所示：

```
<html>
<body>
<form action="XXX" method="POST">
名字：<input type="text" name="first_name">
<br />
姓氏：<input type="text" name="last_name" />
<input type="submit" value="提交" />
</form>
</body>
</html>
```

与通过表单的 Get 方法一样，将 HTML 文件放在指定目录下，通过浏览器访问运行“http://localhost:8080/hello.htm”时，则会出现表单。自行输入姓氏和名字，单击“提交”按钮，基于提交的结果，后台 Servlet 将产生正确的参数数值。

4. Servlet 中复选框数据的传递

复选框用于选择一个以上的数据，如下所示的代码实例，是 HTML 中带有 5 个复选框的表单。

例 6-10 含有复选框的表单。

```
<html>
<body>
<form action="XXXX" method="POST" target="_blank">
<input type="checkbox" name="singing" /> 唱歌
<input type="checkbox" name="dancing" /> 跳舞
<input type="checkbox" name="reading" /> 阅读
```

```
<input type="checkbox" name="traveling" /> 旅游
<input type="checkbox" name="swimming" /> 游泳
<input type="submit" value="选择爱好" />
</form>
</body>
</html>
```

Servlet 程序，处理 Web 浏览器传递的复选框输入的代码如下：

```
out.println(
        "  <li><b>唱歌：</b>: "+ request.getParameter("singing") + "\n"
       +"  <li><b>跳舞：</b>: "+ request.getParameter("dancing") + "\n"
       + "  <li><b>阅读：</b>: "+ request.getParameter("reading") + "\n"
       + "  <li><b>旅游：</b>: "+ request.getParameter("traveling") + "\n"
       + "  <li><b>游泳：</b>: "+ request.getParameter("swimming") + "\n")
```

6.2.4 小节中介绍了各种 Servlet 表单数据传递的方法，在复选框传递数据的方法中，请同学们课后尝试使用循环和枚举的方式，得到选中复选框的参数值，并返回结果。

6.3　创建 HttpServlet

HTTP 是一种基于请求/响应模式的协议，Web 服务器和浏览器通过 HTTP 在 Internet 上发送和接收消息。在 HTTP 中，由客户端建立连接和发送 HTTP 请求，HttpServlet 能够根据 HTTP 请求生成相应的响应结果，前文 6.2.2 小节中已给出有关 HttpServlet 类的介绍。

Servlet 响应 Web 服务器请求的具体步骤如下所述。

（1）Web 客户向 Servlet 容器发出 HTTP 请求。

（2）Servlet 容器创建一个 HttpServletRequest 对象，将 HTTP 的请求信息封装进该对象。

（3）Servlet 容器创建一个 HttpServletResponse 对象。

（4）Servlet 容器调用 HttpServlet 的 service 方法，将 HttpServletRequest 对象和 HttpServletResponse 对象作为 service 方法的参数传给 HttpServlet 对象。

（5）HttpServlet 调用 HttpServletRequest 的相关方法，获取 HTTP 请求信息。

（6）HttpServlet 调用 HttpServletResponse 的相关方法，生成响应数据。

（7）Servlet 容器将 HttpServlet 的响应结果传给 Web 客户。

创建一个 HttpServlet 类，通常涉及以下步骤。

（1）继承 HttpServlet 抽象类。

（2）重载适当的方法，如覆盖 doGet()方法和 doPost()方法。

（3）获取 HTTP 请求信息，可调用 HttpServletRequest 类对象的以下 3 种方法获取：

```
getParameterNames()       //获取请求中所有参数的名字
getParameterValues()       //获取请求中所有参数的值
getParameter()           //获取请求中指定参数的值
```

（4）生成 HTTP 响应结果。HttpServletResponse 对象含有 getWriter()方法，以返回一个 PrintWriter 类对象，PrintWriter 的 print()或 println()方法可以向客户端发送字符串数据流。例 6-11 提供了一个 Servlet 样例。

例 6-11　创建 HttpServlet 的实例。

```
    package mypack;
    import javax.servlet.*;
    import javax.servlet.http.*;
    import java.io.*;
public class HelloServlet extends HttpServlet//第一步：扩展 HttpServlet 抽象类
{
 public void doGet(HttpServletRequest request,
  HttpServletResponse response)throws IOException,ServletException{
//第二步：覆盖 doGet()方法
  String clientName=request.getParameter("clientName");
    if(clientName!=null)
   clientName=new String(clientName.getBytes("ISO-8859-1"),"GB2312");
  else
   clientName="我的朋友"; //第三步：获取 HTTP 请求中的参数信息
  PrintWriter out;//第四步：生成 HTTP 响应结果
  String title="HelloServlet";
  String heading1="HelloServlet 的 doGet 方法的输出：";
  //set content type
  response.setContentType("text/html;charset=GB2312");
  //write html page
  out=response.getWriter();
  out.print("<HTML><HEAD><TITLE>"+title+"</TITLE>");
  out.print("</HEAD><BODY>");
  out.print(heading1);
  out.println("<h1><p>"+clientName+":您好</h1>");
  out.print("</BODY></HTML>");
  out.close();
 }
}
```

在 web.xml 中做如下改动：

```
<servlet>
   <servlet-name>HelloServlet</servlet-name>
   <servlet-class>mypack.HelloServlet</servlet-class>
  </servlet>
  <servlet-mapping>
   <servlet-name>HelloServlet</servlet-name>
   <url-pattern>/hello</url-pattern>
</servlet-mapping>
```

首先在 MyEclipse 中新建一个 web project，命名为 helloServlet。在 src 目录下新建 mypack 包，在包中添加名为 HelloServlet.java 的 Servlet 项目。根据示例，重写关于 doGet 中的代码，并在 WEB-INF 目录下改动 xml 配置文件。

上述 HelloServlet 类扩展 HttpServlet 抽象类，覆盖了 doGet 方法。在 doGet 方法中，通过 getParameter 方法读取 HTTP 请求中的一个参数 clientName。运行程序，并将其部署至 Tomcat 服务器（具体方法参考 1.2.5 小节），然后在浏览器中通过 URL 访问 HelloServlet。在浏览器中输入 http://localhost:8080/helloServlet/hello?clientName=mm，运行结果如图 6.2 所示。

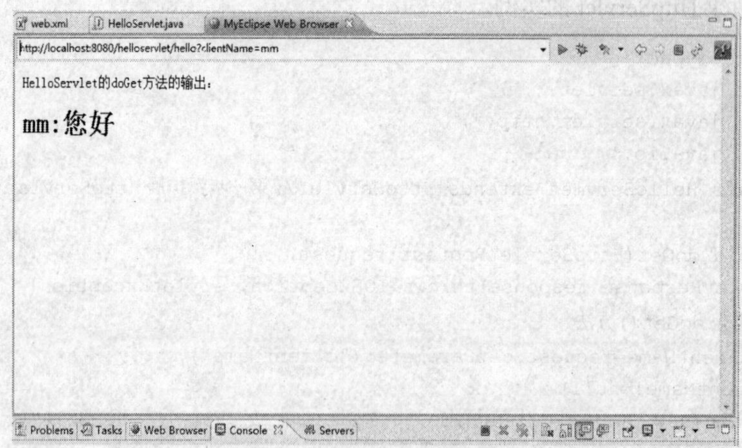

图 6.2　HelloServlet 的执行结果

注意

URL 中 helloServlet 表示 web project 的名称，hello 是在 xml 中定义的 url-pattern。"clientName=mm" 中的参数通过 getParameter 方法被获取并传送，最后生成 HTTP 响应结果。

6.4　JavaBean 技术

6.4.1　JavaBean 基础

JavaBean 是一种可重复使用的、跨平台的软件组件。JavaBean 一般分为可视化组件和非可视化组件两种。可视化组件可以是简单的 GUI 元素，也可以是复杂的组件；非可视化组件用于封装业务逻辑、数据库操作等。其最大的优点在于可以实现代码的可重用性。JSP 访问的通常是后一种 JavaBean。

JavaBean 具有以下特性。

（1）易于维护、使用、编写。

（2）可实现代码的重用性。

（3）可移植性强，但仅限于 Java 工作平台。

（4）便于传输，不限于本地还是网络。

（5）可以以其他部件的模式进行工作。

1. JavaBean 的属性

JavaBean 是具有统一接口格式的 Java 类，由属性(Properties)、方法(Method)、事件(Events)组成，一个 JavaBean 对象的属性应该是可访问的。这个属性可以是任意合法的 Java 数据类型，包括自定义 Java 类。

JavaBean 对象的属性可以是可读可写，或只读，或只写。JavaBean 对象的属性通过 JavaBean 实现类中提供的两个方法来访问：get 方法设置属性，set 方法获取属性。

2. JavaBean 程序实例

例 6-12 中创建了一个名为 Test 的 JavaBean 实例，在 test 中定义了两个属性 username 和 password，

并定义了属性的方法 setXXX()和 getXXX()。其中，以“set”为名称开始的方法是可写的属性，set 方法不应拥有返回类型（即必须为 void），并且只能有一个参数，参数的数据类型必须和属性的数据类型一致。以“get”为名称开始的方法是可读的属性，应返回合适的类型，并且不允许有参数。

例 6-12　创建 JavaBean 的实例。

```
package com;                  //将 Test 这个类放入包 com 中，以便在外部引用
public class Test           //此处为构造函数
{
private String username;
private String password;
        public void setUsername(String username)
{
   this.username=username;    //this 指当前类中的 username,而不是参数 username
}
public void setPassword(String password)
{
   this.password=password;
}
public String getUsername()
{
   return username;
}
public String getPassword()
{
   return password;
}
}
```

3. JavaBean 的访问

<jsp:useBean>标签可以在 JSP 中声明一个 JavaBean，然后使用。声明后，JavaBean 对象就成了脚本变量，可以通过脚本元素或其他自定义标签来访问。<jsp:useBean>标签的语法格式如下：

```
<jsp:useBean id="bean's name" scope="bean's scope" typeSpec/>
```

其中，根据具体情况，scope 的值可以是 page，request，session 或 application。id 值可任意，只要不和同一 JSP 文件中其他<jsp:useBean>中的 id 值一样就行了。

接下来给出的是<jsp:useBean>标签的一个简单用法：

```
<html>
<head>
<title>useBean Example</title>
</head>
<body>

<jsp:useBean id="date" class="java.util.Date" />
<p>The date/time is <%= date %>

</body>
</html>
```

6.4.2　JavaBean 开发模式

JavaBean 有两种开发模式，即 JSP+JavaBean 和 JSP+Servlet+JavaBean。

1. JSP+JavaBean

JSP+JavaBean 两层结构的开发模式比较适用于小型项目，使用 JavaBean 可以把业务处理功能从 JSP 页面分离，使 JSP 页面专注于处理数据的显示，从而使页面代码量减少，以及逻辑清晰。

JSP 从本质上来说，就是把 Java 代码嵌套在静态的 HTML 页面，使 HTML 页面有动态的功能。采用 JavaBean 后，可以在 JSP 网页通过代码访问 JavaBean，也可以通过特定的 JSP 标签访问 JavaBean，后者使得 JSP 网页更接近于 HTML 页面。

例 6-12、例 6-13 为一个典型的用户注册的实例，采用 JSP+JavaBean 模式。JavaBean 传输数据并处理数据，然后在 JSP 显示，该模式的两层分工很明确。

例 6-13 用户注册的实例。

```
login.jsp:
<%@ page language="java" import="java.util.*" pageEncoding="utf-8"%>
<html>
<center>
<form method=post action="welcome.jsp">
username<input type=text name=username>
<br><br>
password<input type=password name=password>
<br><br>
<input type=submit value="注册">
</form>
</center>
</html>
```

welcome.jsp:

```
<%@ page language="java" import="java.util.*" pageEncoding="utf-8"%>
<html>
<jsp:useBean id="hello" class="com.Test" scope="session" />
<jsp:setProperty name="hello" property="*" />
<%
//hello.username = "myname";
%>
your username is:<%= hello.getUsername() %>
<br><br>
your password is:<jsp:getProperty name="hello" property="password"/>
<br><br>
</html>
```

2. JSP+Servlet+JavaBean

使用 JSP+JavaBean 开发模式，当业务逻辑复杂时，JSP 页面的 Java 代码量将变得庞大。若使用一个或多个 Servlet 作为控制器，JavaBean 作为模型处理业务逻辑，JSP 用来显示页面，这样将显示和逻辑分离的模式更适合大型项目。

JSP+Servlet+JavaBean 模型是一种采用基于模型视图控制器（Model+View+Controller）的设计模型，即 MVC 模型（第 7 章将详细介绍），该模型将 JSP 程序的功能分为 3 个层次：模型层、视图层和控制层。JSP+Servlet+JavaBean 模型中，JavaBean 作为模型层，实现各个具体的应用逻辑和功能；Servlet 作为控制层，负责处理 HTTP 请求；JSP 作为视图层，负责生成交互后返回的界面。

例 6-14 JSP+Servlet+JavaBean 实例。

```
package com;
import java.io.*;
import javax.servlet.*;
```

```
import javax.servlet.http.*;
import com.Test;

public class HelloServlet extends HttpServlet
 {
   public void service(HttpServletRequest request,
         HttpServletResponse response)
           throws ServletException, IOException
 {
     response.setContentType("text/html; charset=GBK");
     request.setCharacterEncoding("GBK");
     Test test = new Test();    //创建 JavaBean 的对象
     test.setUserName(request.getParameter("username").toString());
     HttpSession session = request.getSession();
     session.setAttribute("userInfo", userInfo);
    response.sendRedirect("welcome.jsp");  // 前往指定的网页
  }
 }
```

在 web.xml 中做如下改动：

```
<servlet>
     <servlet-name>
        login
     </servlet-name>
     <servlet-class>
        com.HelloServlet
     </servlet-class>
   </servlet>
   <servlet-mapping>
     <servlet-name>login</servlet-name>
     <url-pattern>/login.do</url-pattern>
   </servlet-mapping>
```

最后修改 login.jsp 标签 form 的 action="/login.do"，即实现了一个简单的 MVC 模型。

思考与习题

1. 填空题：JSP 文件中有 5 类元素，分别是模板元素、_____、____、____和动作元素。

2. 填空题：JSP 中的内置对象包括 page 对象、config 对象、___________、___________、out 对象、____________、application 对象、pageContext 对象、exception 对象等。

3. 填空题：Servlet 的生命周期中包含 3 种方法，即 init 方法、_________和 destroy 方法。

4. 填空题：HTTP 是一种__________的协议，Web 服务器和浏览器通过 HTTP 在 Internet 上发送和接收消息。

5. 填空题：JavaBean 是具有统一接口格式的 Java 类，由_______、_______和______组成。

6. 简答题：简述 session 对象与 application 对象的区别。

7. 简答题：简述 Servlet 的任务和生命周期。

8. 编程题：尝试使用循环和枚举的方式，得到选中复选框的参数值，并返回结果。

9. 编程题：采用 JSP+Servlet+JavaBean 的模式，实现简单的注册系统，并通过上章介绍的 JDBC 知识，将该注册系统连接数据库。

第 7 章 MVC 模式和 Struts2 框架

Struts2 是基于 MVC 架构的框架，是 Struts 的下一代产品。Struts2 框架是在 Struts1 和 WebWork 的技术基础上进行合并的全新的框架。而 MVC 则是一种架构型模式，本身并不引入新的功能，只是用来改善应用程序的架构，使得应用的模型和视图相分离，从而达到更好的开发和维护效率。

本章将首先介绍 MVC 模式基础，然后详细介绍 Struts2 框架的安装配置、工作流程和编程实例等相关知识。

7.1 MVC 模式基础

7.1.1 MVC 模式简介

MVC 是一种优秀的软件设计模式，它将应用分成模型层、视图层和控制层 3 个层次，从而使同一个应用程序使用不同的表现形式。自从 MVC 提出以后，出现了许多 MVC 框架，其中 Struts 是第一个使用 MVC 架构的框架，后来从 Struts 发展而成的 Struts2 集成了 Struts 和 WebWork 两个框架的优点，使其具有更好的扩展性、更强大的功能，成为了目前主流的 MVC 框架。

MVC 是 Xerox PARC 在 20 世纪 80 年代为编程语言 Smalltalk-80 发明的，MVC 框架如图 7.1 所示，其中 M 表示数据模型，V 是视图界面，C 是指控制器。使用 MVC 的目的是将数据模型和视图界面实现代码分离。

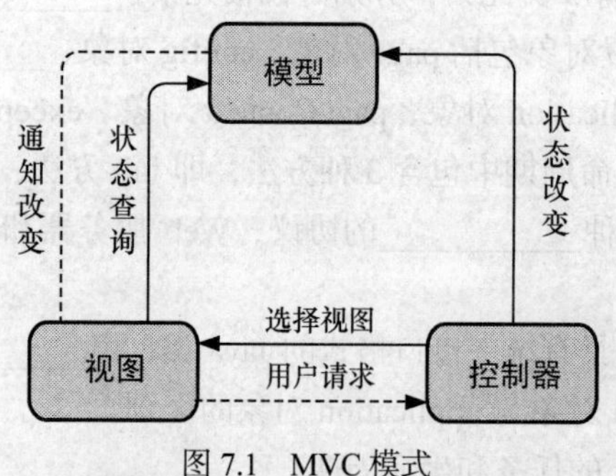

图 7.1　MVC 模式

MVC 模型是一种交互界面的结构组织模型，它能够使软件的计算模型独立于界面的构成。MVC 模型最早应用在 SmallTalk-80 的环境中时，是许多交互和界面系统的构成基础。后来 Microsoft 的 MFC 模型基础类也遵循了 MVC 的设计思路。MVC 设计模式更深层次地影响了软件开发人员的任务分工，使软件开发更加方便。MVC 模式的原理如图 7.2 所示。

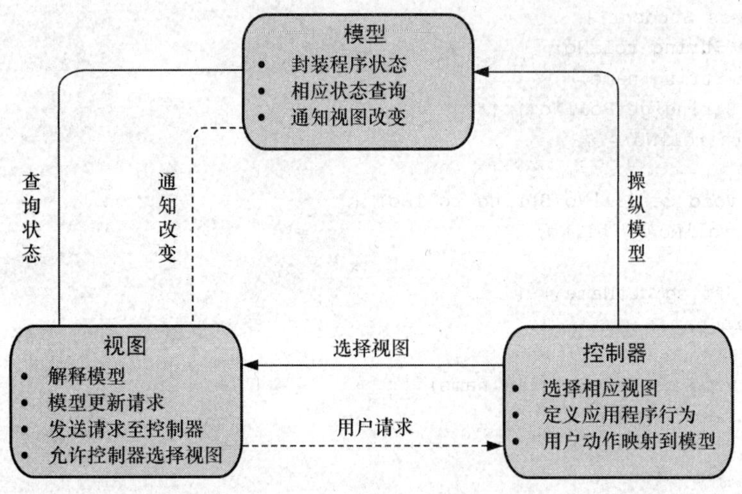

图 7.2　MVC 原理图

7.1.2　模型、视图和控制器

MVC 模式代表 Model-View-Controller（模型-视图-控制器）模式，各模式的作用如下。

- Model（模型）：模型代表一个存取数据的对象或 Java POJO。它也可以带有逻辑，在数据变化时更新控制器。
- View（视图）：视图代表模型包含的数据的可视化。
- Controller（控制器）：控制器作用于模型和视图上。它控制数据流向模型对象，并在数据变化时更新视图，它将视图与模型互相分离。

模型、视图和控制器的分离，使得一个模型可以具有多个显示视图。如果用户通过某个视图的控制器改变了模型的数据，所有其他依赖于这些数据的视图都会反映出这些变化。因此，当数据发生变化时，控制器将会通知所有相关的视图，导致显示的更新。

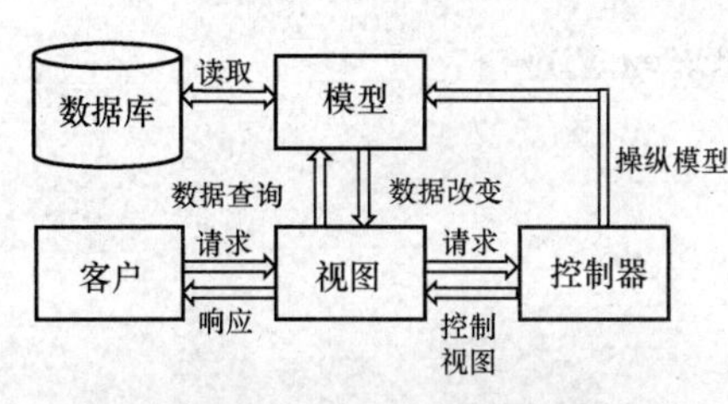

图 7.3　MVC 模块作用图

MVC 各模块的相互作用如图 7.3 所示，客户可以从视图提供的客户界面上浏览数据或发出请求，客户的请求将交给控制器处理。控制器根据不同的客户请求调用不同的模型方法，完成数据的更新，然后调用视图的方法并将响应的结果返回给客户。视图也可以直接访问模型，查询数据的信息。当模型中的数据发生变化时，它会通知视图刷新界面，显示更新后的数据。

7.1.3　MVC 的实现

下面我们将实现一个简单的 MVC 模型实例。创建一个 Student 对象作为模型，StudentView 则是一个视图类，负责把学生详细信息输出到控制台。StudentController 是控制器，负责存储数据到 Student 对象中，并相应地更新视图 StudentView。

1. 创建模型

首先创建一个作为模型的 Student 类对象。

例 7-1 创建 Student 类对象模型。

```
public class Student {
   private String rollNo;
   private String name;
   public String getRollNo() {
      return rollNo;
   }
   public void getRollNo(String rollNo) {
      this.rollNo = rollNo;
   }
   public String getName() {
      return name;
   }
   public void getName(String name) {
      this.name = name;
   }
}
```

2. 创建视图

例 7-2 创建 StudentView 视图。

```
public class StudentView {
   public void printStudentDetails(String studentName, String studentRollNo){
      System.out.println("Student: ");
      System.out.println("Name: " + studentName);
      System.out.println("Roll No: " + studentRollNo);
   }
}
```

3. 创建控制器

例 7-3 创建 StudentController 控制器。

```
public class StudentController {
   private Student model;
   private StudentView view;

   public StudentController(Student model, StudentView view){
      this.model = model;
      this.view = view;
   }
   public void setStudentName(String name){
      model.getName(name);
   }
   public String setStudentName(){
      return model.getName();
   }
   public void setStudentRollNo(String rollNo){
      model.getRollNo(rollNo);
   }
   public String setStudentRollNo(){
      return model.getRollNo();
   }
   public void updateView(){
      view.printStudentDetails(model.getName(), model.getRollNo());
```

```
    }
}
```

4. 演示 MVC 模式

使用 StudentController 方法来演示 MVC 设计模式的用法。

例 7-4　演示 MVC 设计模式的用法。

```
public class MVCPatternDemo {
   public static void main(String[] args) {
    Student model  = retriveStudentFromDatabase();              //从数据中获取学生记录
    StudentView view = new StudentView();        //创建一个视图：把学生详细信息输出到控制台
    StudentController controller = new StudentController(model, view);
    controller.updateView();
    controller.setStudentName("John");
    controller.updateView();
     }                                                                  //更新模型数据
   private static Student retriveStudentFromDatabase(){
      Student student = new Student();
      student.getName("Robert");
      student.getRollNo("10");
      return student;
   }
}
```

以上代码实现了一个简单的 MVC 模型实例，运行实例验证，最终会得到以下结果：

```
Student:
Name: Robert
Roll No: 10
Student:
Name: John
Roll No: 10
```

7.2　Struts2 框架基础

在 MVC 模型逐渐流行后，出现了各种基于 MVC 架构的框架，Struts2 就是其中之一。Struts2 的体系结构与 Struts1 的体系结构差别巨大。因为它以 WebWork 为核心，采用拦截器的机制来处理用户的请求，所以从一定的程度上 Struts2 可以理解为 WebWork 的更新产品。虽然从 Struts1 到 Struts2 有着较大变化，但是相对于 WebWork，Struts2 的变化很小。

7.2.1　Struts2 概述

作为 WebWork 框架和 Struts 联手打造的框架，Struts2 具有良好的功能。

（1）POJO 格式和 POJO 动作：Struts2 可以使用任何的 POJO 可接收的形式输入，同样，也可以得到任何 POJO 的 Action 类。

（2）标签支持：Struts2 的新标签可以让程序代码更加有效率。

（3）Ajax 支持：Struts2 中运用了 Web2.0 技术，并支持创建 Ajax 的标签。

（4）易整合：Struts2 与其他框架如 Spring、Tiles 和 SiteMesh 等更容易集成与整合。

（5）模板与插件支持：Struts2 支持生成使用模板，并且 Struts2 可以加强和扩大使用插件。

（6）概要分析：Struts2 可以集成配置、调试和分析应用程序。

（7）显示技术：Struts2 含有多个支持视图的选项（JSP，Freemarker，Velocity 和 XSLT）。

7.2.2 Struts2 工作流程

Struts2 框架是一个表示层的框架，主要用于处理应用程序与客户端交互问题。图 7.4 所示为 Struts2 的基本工作流程。

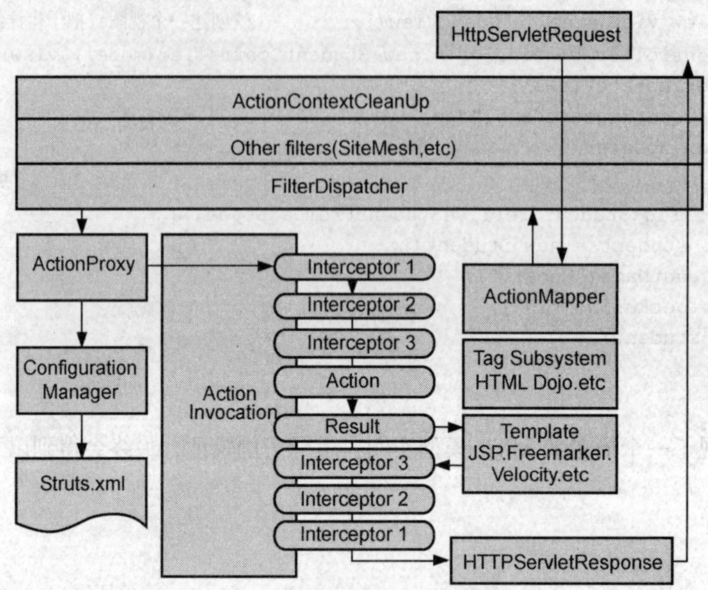

图 7.4 Struts2 框架工作流程图

（1）客户端向 Web 容器（如 Tomcat）发送请求。

（2）过滤器将请求过滤（如 ActionContextCleanUp 过滤器等），并传递给 FilterDispatcher。

（3）FilterDispatcher 收到请求信息后，根据 URL 在 ActionMapper 中搜索指定 Action 的映射信息。

（4）如果找到符合的映射信息，ActionProxy 将通过 Configuration Manager 询问框架的配置文件"struts.xml"，搜索被请求的 Action 类。

（5）ActionProxy 创建一个被请求 Action 的实例，该实例将用来处理请求信息。

（6）如果在"struts.xml"文件中存在与被请求 Action 相关的拦截器配置，那么该 Action 的实例被调用的前后，这些拦截器也会先被执行。

（7）Action 对请求处理完毕后，ActionInvocation 负责根据结果码字符串在 struts.xml 的配置中找到对应的返回结果（JSP、Velocity 模板、FreeMarker 模板等），并返回给客户端。

（8）拦截器被再次执行。

（9）过滤器被再次执行。

Struts2 中核心的类与接口有以下各项。

ActionMapper：根据请求的 URI 查找是否存在对应 Action 调用。

ActionMapping：保存调用 Action 的映射信息，如 namespace、name 等。

ActionProxy：在 XWork 和真正的 Action 之间充当代理。

ActionInvocation：表示 Action 的执行状态，保存拦截器、Action 实例。

Interceptor：可以在请求处理之前或者之后执行的 Struts2 组件，Struts2 绝大多数功能通过拦截器完成。

7.2.3　Struts2 配置文件

当 Struts2 创建系统的 Action 代理时，需要使用 Struts2 的配置文件。Struts2 的配置文件有两份，配置 Action 的“struts.xml”和配置 Struts2 全局属性的“struts.properties”文件。

1. struts.xml 的配置

“struts.xml”是 Struts2 中的核心配置文件，文件中定义了 Struts2 的系列 Action，并指定该 Action 的实现类，定义其处理结果和视图资源之间的映射关系。“struts.xml”通常放在 WEB-INF/classes 目录下，在该目录下的 struts.xml 文件可以被自动加载，也可以放到 src 目录下，struts.xml 中包含如下几个主要的配置元素，示例代码如下：

```
<struts>
  <include file=" " />
  <constant name="" value=""/>
   <package name="" namespace="" extends="">
      <action name="" class="">
        <result name=""></result>
      </action>
    </package>
     </struts>
```

（1）<include>元素

利用 include 标签，可以将一个 struts.xml 配置文件分割成多个配置文件，然后在 struts.xml 中使用<include>标签引入其他配置文件。

比如一个网上购物程序，可以把用户配置、商品配置、订单配置分别放在 3 个配置文件 user.xml、goods.xml 和 order.xml 中，然后在 struts.xml 中将以下这 3 个配置文件引入。以下给出 struts.xml、user.xml 文件的代码实例，goods.xml 和 order.xml 与 user.xml 类似，这里不再详细列举。

struts.xml：

```
<?xml version="1.0" encoding="UTF-8"?>
    <!DOCTYPE struts PUBLIC
        "-//Apache Software Foundation//DTD Struts Configuration 2.0//EN"
    "http://struts.apache.org/dtds/struts-2.0.dtd">
<struts>
    <include file="user.xml"/>
    <include file="goods.xml"/>
    <include file="order.xml"/>
</struts>
```

user.xml：

```
<?xml version="1.0" encoding="UTF-8"?>
<!DOCTYPE struts PUBLIC
    "-//Apache Software Foundation//DTD Struts Configuration 2.0//EN"
    "http://struts.apache.org/dtds/struts-2.0.dtd">
<struts>
    <package name="wwfy" extends="struts-default">
        <action name="login" class="wwfy.user.LoginAction">
```

```
            <!--省略 Action 其他配置-->
        </action>
        <action name="logout" class="wwfy.user.LogoutAction">
            <!--省略 Action 其他配置-->
        </action>
    </package>
</struts>
```

（2）<contant>元素

<contant>元素用于配置常量信息，如开发模式、字符集编码等，可以改变 Struts2 框架的一些行为。其中 name 属性表示常量名称，value 属性表示常量值。下面是使用<contant>元素的示例代码：

```
<?xml version="1.0" encoding="UTF-8"?>
<!DOCTYPE struts PUBLIC
    "-//Apache Software Foundation//DTD Struts Configuration 2.0//EN"
    "http://struts.apache.org/dtds/struts-2.0.dtd">
<struts>
    <constant name="struts.devMode" value="true"/>              //设置开发模式
    <constant name="struts.i18n.encoding" value="GB2312"/>    //设置编码形式为 GB2312
    <!--省略其他配置信息-->
</struts>
```

（3）<package>元素

在 Struts2 框架中是通过包来管理 action、result、interceptor、interceptor-stack 等配置信息的。包提供了将多个 action 组织为一个模块的方式，并且可以继承已定义的包，这样有利于简化维护工作，提高代码的重用性。<package>元素的属性如表 7.1 所示。

表 7.1　<package>元素的属性

属　性	是 否 必 需	描　　述
name	是	包名，作为其他包应用本包的标记
extends	否	设置本包继承的其他包
namespace	否	设置包的命名空间
abstract	否	设置为抽象包

name：包名，作为其他包引用本包的标识符，该属性是必须的。

extends：用于继承其他包，该属性是可选的。当一个包通过配置 extends 属性继承了另一个包的时候，该包将会继承父包中所有的配置，包括 action、result、interceptor 等。由于包信息的获取是按照配置文件的先后顺序进行的，所以父包必须在子包之前被定义。通常我们配置 struts.xml 的时候，都继承一个名为“struts-default.xml”的包，该包是一个内置包，配置了所有的内置结果类型。struts-default 包在 struts-default.xml 文件中定义，而 struts-default.xml 是 Struts2 默认配置文件，会自动加载。

namespace：用于设置命名空间，该属性是可选的。主要是针对大型项目中 action 的管理，更重要的是解决 action 重名问题，因为不同命名空间的 action 可以使用相同的 action 名。

abstract：设置为抽象包，该属性是可选的。abstract 抽象包的包名是唯一的，且如果有一个包被设置为抽象包，则该包中不可包含 action 的配置信息，但可以被其他包继承。

（4）<action>元素

顾名思义，<action>元素是用来配置 Action 信息的，<action>元素的属性如表 7.2 所示。

表 7.2　<action>元素的属性

属　　性	是 否 必 需	描　　述
name	是	请求的 Action 名称
class	否	Action 处理类对应具体路径
method	否	指定 Action 中的方法名
converter	否	指定 Action 使用的类型转换器

如果没有为 Dction 指定 result，默认值为 success。如果不为 Action 设置 class，则 Action 默认是 ActionSupport。如果不为 Action 设置 method 属性，默认是执行 Action 中的 execute 方法。但是一个 Action 也有可能要处理多个业务逻辑，此时需要在该 Action 类中定义多个方法，每一个不同的请求需要转交给不同的方法去做处理，这时就需要使用 method 属性。其格式如下所示：

```
<action name="loginUser"
    class="cn.jbit.houserent.action.UserAction" method="login">
    <result>/page/manage.jsp</result>
    <result name="input">/page/login.jsp</result>
    <result name="error">/page/error.jsp</result>
</action>
<action name="registerUser"
        class="cn.jbit.houserent.action.UserAction" method="register">
        <result>/page/success.jsp</result>
        <result name="input">/page/register.jsp</result>
        <result name="error">/page/error.jsp</result>
</action>
```

Struts2 在根据<action>元素的 method 属性查找执行方法时有两种途径：查找与 method 属性值完全一致的方法，以及查找 doMethod()形式的方法。例如在上例中，当请求/login.action 时，Struts2 会首先在 UserAction 中根据 method 属性查找 login()方法；如果找不到，则会查找 doLogin()方法。

使用 method 优点是可以减少 Action 类的数目，缺点是由于要为同一个 Action 配置多个名称，配置文件中将产生大量的冗余代码。为了消除这种冗余现象，Struts2 提供了通配符的方式。以下是改进后的代码片段：

```
<action name= "*User"
    class="cn.jbit.houserent.action.UserAction" method="{1}">
    <result>/page/{1}_success.jsp</result>
    <result name="input">/page/{1}.jsp</result>
    <result name="error">/page/error.jsp</result>
</action>
```

在上述代码中，<action>元素的 name 属性值为“*User”，其中的“*”即为通配符，它表示所有以“User.action”结尾的请求都会交给 Action 做处理，如“loginUser.action”和“registerUser.action”。method 属性值为“{1}”，其中的“{1}”表示 name 属性中第一个“*”值。例如，如果请求为“loginUser.action”，那么“{1}”代表“login”，则 method 属性值为“login”，此时会调用 Action 的 login 方法。

（5）<result>元素

Struts2 通过在 struts.xml 文件中使用<result>元素来配置请求返回的结果页。<result>元素的配置由两部分组成：一部分是 result 所代表的实际资源的位置及 result 名称；另一部分就是 result 的类型，由 result 元素的 type 属性进行设定。

除了上述的 5 个元素外，“Struts2.xml”中还有其他的配置元素，如表 7.3 所示，这里不一一详细说明。

表 7.3　Struts2.xml 中配置元素及其属性

配置元素名称	描　述
include	引入其他 xml 配置文件
contant	配置常量信息
bean	由容器创建并注入的组件
package	包含一系列 Action 及拦截器信息，并对其进行统一管理
default-action-ref	配置默认 Action
default-class-ref	配置默认 class
default-interceptor-ref	配置默认拦截器，对包含范围内所有 Action 有效
global-results	配置全局结果集，对包含范围内所有 Action 有效
global-exception-mappings	配置全局异常映射，对包含范围内所有 Action 有效
result-types	配置自定义返回结果类型
interceptors	包含一系列拦截器配置信息
action	包含与 Action 操作相关的一系列配置信息
exception-mappings	配置异常映射，Action 范围内有效
interceptor-ref	配置 Action 应用的拦截器
result	配置 Action 的结果映射

2. struts.properties 的配置

“struts.properties”文件的形式是 key-value 键值对，它指定了 Struts2 应用的全局属性，该文件的示例如下：

```
#指定 Struts2 处于开发状态
struts.devMode = false
#指定当 Struts2 文件配置改变后，Web 框架是否重新加载配置文件
struts.configuration.xml.reload=true
```

7.2.4 Struts2 标签库

Struts2 的标签库也是 Struts2 的重要组成部分，标签库提供了非常丰富的功能，如表现层数据处理的功能，以及基本的流程控制功能，还提供了国际化、Ajax 支持等功能，在一定程度上可以提高应用程序的开发效率。

1. 表单标签

下面的 JSP 页面的表单定义片段则使用了 Struts2 表单标签生成表单元素：

```
<!--使用 Struts 2 标签定义一个表单 -->
<s:form method="post" action="basicvalid.action">
<!-- 定义 3 个表单域-->
    <s:textfield label="姓名" name="name"/>
    <s:textfield label="年龄" name="age"/>
    <s:textfield label="喜欢的颜色" name="answer"/>
<!-- 定义一个提交按钮-->
    <s:submit/>
</s:form>
```

Struts2 中不仅提供了与 HTML 表单标签作用相同的标签，如上例所示，也提供了自己的表单标签，其页面代码较 HTML 表单标签更加简洁。

2. 控制标签

Struts2 标签库功能非常强大，除了表单标签外，还有控制标签，主要用于控制输入流程及访问值栈中的值。

if/elseif/else 标签用于完成分支控制，其拥有一个 test 属性，test 中表达式的值用来决定标签里内容是否显示，用法如下：

```
<s:if test="#request.username=='mm'">欢迎 mm</s:if>
<s:elseif test="#request.username=='cc'">不欢迎 cc</s:if>
```

iterator 标签用于遍历集合(java.util.Collection)或者枚举值(java.util.Iterator)类型的对象，value 属性表示集合或枚举对象，status 属性表示当前循环的对象，在循环体内部可以引用该对象的属性。用法如下：

```
<s:iterator value="userList" status="user">
姓名：<s:property value="user.userName" />
年龄：<s:property value="user.age" />
</s:iterator>
```

3. 数据标签

数据输出标签可以输出页面中的元素、属性、隐含变量等，既包括静态文本的输出，也包括 Struts2 集成的各种变量的输出。

（1）JavaBean 标签：用于创建一个 JavaBean 对象，name 表示 JavaBean 类全名，var 表示变量的实例名，并可以包含<s:param>设置实例化输入参数，用法如下：

```
<s:bean name="xxx,xxx,xxx" var="xxx">
<s:param name="name" value="mm" />        //为该 JavaBean 的构造方法传递参数
</s:bean>
```

（2）传递参数标签：用来为其他标签提供参数，所以一般是嵌套在其他标签的内部，name 属性用来指定参数名称，value（可选）属性用来指定参数值，有两种用法：

```
<s:param name="username" value="mm" />     //这种方式的参数值会以 String 格式放入 stack
<s:param name="username">mm</s:param>      //这种方式的参数值会以 Object 的格式放入 stack
```

（3）页面标签：用来包含一个 Servlet 的输出（Servlet 或 JSP 页面），使用 value 属性指定包含页名称，可以使用<s:param>传递参数给子页，用法如下：

```
<s:include value="/test.jsp">
<s:param name="username">mm</s:param>              //传递 username=mm 给 test.jsp 页面
</s:include>
```

（4）值栈标签：用来加入一个值到值栈中，用法如下：

```
<s:push value="user">
<s:property value="name" />                          //为 user 指定 name 属性
<s:property value="age" />                           //为 user 指定 age 属性
</s:push>
```

（5）属性标签：用于输出 value 属性的值，并拥有一个 default 属性，在 value 对象不存在时显示。escape 属性为 true，来输出原始的 HTML 文本，用法如下：

```
<s:property value="mm">
```

如果值栈中存在一个名为“mm”的对象，那么页面被请求时，该对象在值栈中的值将被输出，否则输出 mm 字符串。

（6）国际化标签：用于输出国际化信息，name 属性指定国际化资源文件中消息文本的 key，可以与<s:i18n>标签结合使用。该标签作用范围内，指定的国际化资源优先级大于当前 locale 对应的国际化优先级。用法如下：

```
<s:i18n name="application_zh_CN">
输出汉语信息：<s:text name="message"/><br>
</s:i18n>
输出英文信息：<s:text name="message"/>
```

上述代码中，当使用 i18n 标签时，在该标签内使用 name 指定的国际化资源，在标签外使用系统默认的国际化资源。

7.3 Struts2 实现的 MVC 模式

7.3.1 Struts2 架构

Struts2 不同于传统的 MVC 框架。从架构上看，属于上拉 MVC 框架，图 7.5 为 Struts2 的架构图。在 Struts2 的 Model-View-Controller 模式实现以下 4 个核心组件：

（1）动作-Actions；

（2）拦截器-Interceptors；

（3）值栈/OGNL；

（4）结果与视图。

图 7.5 描述了 Struts2 的体系结构模型、视图和控制器。其中，Dispatcher Filter 是核心过滤器。模型（model）中的核心是 Action，Action 是一个简单的 POJO 对象，Action 的开发是基于 Struts2 的项目的核心任务。Interceptor（拦截器）是一个遵守 Struts2 规范的对象，控制层会在 Action 方法执行之前及之后执行 interceptor 中的代码。值栈和 OGNL 提供了一个通用的路线，用于链接和集成其他组件。Struts2 框架通过配置文件中的 result 属性，将 Action 和视图资源相联系，Result 对应的视图资源，即为 MVC 模式中的视图部分（View）。

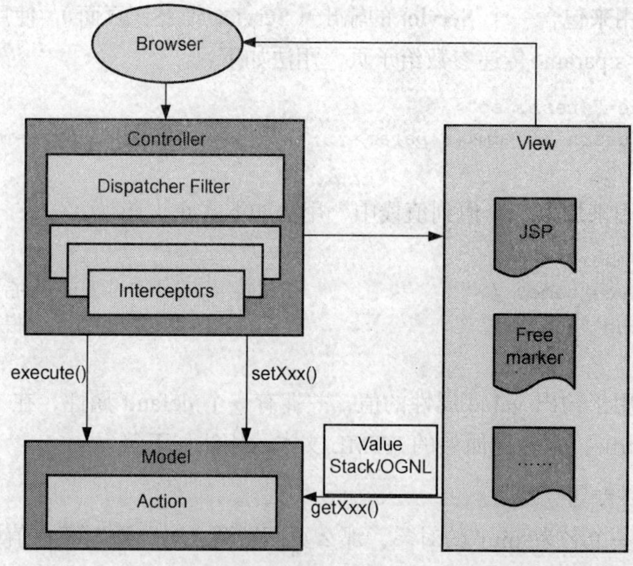

图 7.5　Struts2 架构图

请求的生命周期如下所述。

（1）用户将请求发送至服务器。

（2）Dispatcher Filter 接收到请求信息后，搜索指定 Action 的映射信息。

（3）ActionProxy 在配置文件“struts.xml”中搜索请求的 Action 类，并创建一个 Action 实例。

（4）此时如果请求的 Action 在“struts.xml”中有拦截器的配置信息，那么 Action 实例的调用将触发拦截器的执行。一般配置拦截器的功能适用于验证、文件上传等。

（5）执行 Action 请求的操作。

（6）请求处理完毕后返回一个逻辑视图，由逻辑视图寻找物理视图，最后的结果由视图准备，并且将结果返回给用户。

在随后的章节中，将详细地介绍 Struts2 每个组件和“struts.xml”中的配置信息。

7.3.2　FilterDispatcher 核心过滤器

org.apache.Struts2.dispatcher.FilterDispatcher 是 Struts2 中的核心过滤器。用户提交请求后，请求被 web.xml 中定义的过滤器 FilterDispatcher 拦截，在 FilterDispatcher 中主要经过 3 层过滤器的处理，分别是 ActionContext CleanUp、其他过滤器（Othter Filters、SiteMesh 等）和 FilterDispatcher。

在 FilterDispatcher 过滤器中首先询问 ActionMapper 是否需要调用某个 Action 来处理请求，如果 ActionMapper 决定需要调用某个 Action，FilterDispatcher 则把请求的处理交给 ActionProxy，ActionProxy 通过配置文件“struts.xml”找到需要调用的 Action 类，然后 ActionProxy 创建一个 ActionInvocation 实例，并调用该 Action，但在调用之前，ActionInvocation 会根据配置加载 Action 相关的所有 Interceptor，等 Action 执行完毕，ActionInvocation 负责根据“struts.xml”中的配置找到对应的返回结果 result。

在介绍 FilterDispatcher 之前，首先介绍过滤器的概念。

过滤器提供一种面向对象的模块化机制，用于将公共任务封装到可插入的组件中，这些组件通过一个配置文件来声明并动态地处理。实现过滤器一般需要 3 个步骤：首先编写过滤器实现类的程序，然后把该过滤器添加到 web.xml 中声明，最后把过滤器与应用程序一起打包并部署。

过滤器 API 一共包含 3 个接口：Filter、FilterChain 和 FilterConfig。其中在 Filter 接口中，讲解如下。

init()：该方法在实例化过滤器时调用，主要用于使过滤器为处理做准备。容器为这个方法传递一个包含有配置信息的 FilterConfig。

doFilter()：过滤器只拥有一个用于处理请求和响应的方法 doFilter()。

destroy()：该方法由容器在销毁过滤器实例之前调用，以便能够执行任何必需的清理代码。

FilterDispatcher 也遵循一般过滤器的规则，实现了 init()、doFilter()、destroy()等 3 个接口，在 init()接口里主要实现创建 Dispatcher 和设置默认包的功能。

```
public void init(FilterConfig filterConfig) throws ServletException {
  try {
    this.filterConfig = filterConfig;
    initLogging();                                    //初始化日志
    Dispatcher = createDispatcher(filterConfig);      //创建一个 Dispatcher
    dispatcher.init();
    dispatcher.getContainer().inject(this);           //设定默认包
    staticResourceLoader.setHostConfig(new FilterHostConfig(filterConfig));
    }
```

```
    finally {
      ActionContext.setContext(null);
    }
  }
```

destroy()接口里主要实现了销毁 Dispatcher 和上下文的功能。

```
public void destroy() {
  if (dispatcher == null) {                               //如果 Dispatcher 为空，则正常结束
    log.warn("something is seriously wrong, Dispatcher is not initialized (null) ");
  }
  else {
    try {                                                 //销毁 Dispatcher
      dispatcher.cleanup();
    }
    finally {
    ActionContext.setContext(null);
    }
  }
  }
```

doFilter()接口里主要实现处理请求和响应的方法。

```
public void doFilter(ServletRequest req, ServletResponse res, FilterChain chain)
throws IOException, ServletException {
HttpServletRequest request = (HttpServletRequest) req;          //获取用户请求的 request
HttpServletResponse response = (HttpServletResponse) res;
ServletContext servletContext = getServletContext();
String timerKey = "FilterDispatcher_doFilter: ";                        //加上时间戳
try {
  ValueStack stack = dispatcher.getContainer().getInstance(ValueStackFactory.class).
     createValueStack();                                              //设定上下文或栈
  ActionContext ctx = new ActionContext(stack.getContext());
  ActionContext.setContext(ctx);
  //重新封装 request，记录使用的语言、编码方式、是否是上传文件等
  UtilTimerStack.push(timerKey);
  request = prepareDispatcherAndWrapRequest(request, response);
  ActionMapping mapping;                                          //获取 ActionMapping
  try {
    mapping = actionMapper.getMapping(request,
    dispatcher.getConfigurationManager());
  }
  catch (Exception ex) {                    //如果没有找到合适的 ActionMapping，则抛出异常
  dispatcher.sendError(request, response, servletContext,HttpServletResponse.
  SC_INTERNAL_SERVER_ERROR, ex);
  return;
  }
  if (mapping == null) {              //如果没有配置 ActionMapping，则判断是否为静态资源
    String resourcePath = RequestUtils.getServletPath(request);   //获取访问请求的路径
    if ("".equals(resourcePath) && null != request.getPathInfo()) {
      resourcePath = request.getPathInfo();
    }
    if (staticResourceLoader.canHandle(resourcePath)) {          //判断是否为静态资源
      staticResourceLoader.findStaticResource(resourcePath, request, response);
    }
  else {                                              // 如果不是，则继续执行下一个过滤器
    chain.doFilter(request, response);
  }
```

```
    return;
  }
  dispatcher.serviceAction(request, response, servletContext, mapping);
                                          //在正常情况下，调用 serviceAction 方法
}
finally {
  try {
    ActionContextCleanUp.cleanUp(req);          //清空上下文和时间戳
  }
  finally {
    UtilTimerStack.pop(timerKey);
  }
}
}
```

7.3.3 Action 详解

在 Struts2 中，Action 可以以多种形式存在：实现 Action 接口、继承 ActionSupport 和多方法的 Action。

1. 实现接口的 Action

Struts2 的 Action 属于 MVC 模型层，Action 中的方法代表业务逻辑，Action 中的属性代表请求中的参数。当 Action 类实现 Action 接口后，开发人员可以使用接口提供的常量以便项目的统一管理，使得 Action 类更具有规范性。

Action 接口的源代码如下：

```
public abstract interface com.opensymphony.xwork2.Action{
 public static final java.lang.String SUCCESS = "success";      // 定义常量
 public static final java.lang.String NONE = "none";
 public static final java.lang.String ERROR = "error";
 public static final java.lang.String INPUT = "input";
 public static final java.lang.String LOGIN = "login";
 public abstract java.lang.String execute() throws Exception;   // 定义抽象方法
 }
```

下面是一个实现 Action 接口的 Action 类的示例代码，示例中传输参数 username 和 userpass，由 Action 类实现该接口。

例 7-5　实现 Action 接口的 Action 类的示例代码。

```
package action;
import com.opensymphony.xwork2.ActionSupport;
import com.opensymphony.xwork2.ModelDriven;
public class UserAction implements Action{
  private String username;
  private String userpass;
  public String getUsername() {
    return username;
  }
  public void setUsername(String username) {
    this.username = username;
  }
  public String getUserpass() {
    return userpass;
  }
  public void setUserpass(String userpass) {
```

```
    this.userpass = userpass;
  }
  public String execute() throws Exception {    //省略具体的逻辑业务处理
    return SUCCESS;
  }
}
```

2. 继承 ActionSupport 实现 Action

Struts2 除了提供 Action 接口外，还提供了一个名为“com.opensymphony.xwork2.ActionSupport”的类，该类实现了 Action 接口、Validateable 接口、ValidationAware 接口、TextProvider 接口和 LocaleProvider 接口。通过继承 ActionSupport 来实现 Action 是推荐的做法，因为 ActionSupport 中提供了输入验证、访问国际化资源等常用方法，使得编写 Action 时代码较为简单。例 7-6 是一个简单的继承 ActionSuppoort 类的 Action 类的示例代码。

例 7-6 继承 ActionSuppoort 类的 Action 类的示例代码。

```
package mypack;
import com.opensymphony.xwork2.ActionSupport;
public class HelloAction extends ActionSupport {
  private String message;                          //封装 HTTP 请求参数的属性
  public String getMessage() {                     //属性的 get 方法
    return message;
  }
  public void setMessage(String message) {         //属性的 set 方法
    this.message = message;
  }
  public String execute() throws Exception {
    message = "Hello Word";
    return SUCCESS;
  }
  public Locale getLocale() {                      //获取语言/地区信息
    return super.getLocale();
  }
  public String getText(String key, String defaultValue, List args,
        ValueStack stack){                         //获取国际化信息
    return super.getText(key, defaultValue, args, stack);
  }
  public ResourceBunle getTexts() {                //访问国际化资源包
    return super.getTexts();
  }
  ......
}
```

3. 多方法的 Action

多方法的 Action 中，一个 Action 对应多个业务逻辑，却只写一个 Action 来实现，struts.xml 的配置也相对简单，Action 中自定义方法的声明格式如下：

```
public String 方法名() throws Exception{}
```

例 7-7 是以用户表 Userinfo 的常用操作为例的一段 Action 代码。

例 7-7 Action 中常用操作代码示例。

```
package action;
import po.Userinfo;
```

```
import service.UserService;
public class  CrudUserAction  extends  ActionSupport
implements  ModelDriven<Userinfo> {                         // crud 业务方法
   private UserService userservice = new UserService();
   private  Userinfo userinfo=new Userinfo();               // 模型对象 userinfo
   public Userinfo getModel() {                             // 模型对象 userinfo
      return userinfo;
}
public String create() throws Exception {                   // 增加
   userservice.createUser(userinfo);
   return SUCCESS;
}
public String retrive() throws Exception {                  // 查询
ActionContext.getContext().put("userlist", userservice.selectUsers());
// 查询结果放在 request 中
   return "list";
}
public String update() throws Exception {                   //修改
   userservice.updateUser(userinfo);
   return SUCCESS;
}
public String delete() throws Exception {
   userservice.deleteUser(userinfo.getUsername());          //删除
   return SUCCESS;
}
public String execute() throws Exception {                  // 默认的 execute
   return SUCCESS;
}
}
```

struts.xml 中的部分代码如下：

```
<struts>
  <package name="actions" extends="struts-default">
  <action name="CrudUser" class="action.CrudUserAction">
     <result>/success.jsp</result>
     <result name="list">/UserList.jsp</result>
       </action>
  </package>
</struts>
```

7.3.4　值栈与 OGNL 表达式

1. 值栈

值栈是 Struts2 中一个重要的概念，几乎所有的 Struts2 操作都需要同值栈打交道。值栈的含义正如它的名字所表示的那样——对象所组成的栈。OGNL 的全称是 Object-Graph Navigational Language（对象图导航语言），提供了访问值栈中对象的统一方式。它是一种功能强大的表达式语言（Expression Language，EL），通过它简单一致的表达式语法，可以存取对象的任意属性、调用对象的方法、遍历整个对象的结构图、实现字段类型转化等功能。它使用相同的表达式去存取对象的属性。

值栈中的对象构成及其排列顺序如下所述。

（1）临时对象：该对象是在程序执行过程中，由容器自动创建并存储到值栈中的。临时对象的值并不固定，会随着应用的不同而变化。举个例子来说，像 JSP 标签所遍历的对象容器中，当前访问到的值就是临时对象。

（2）模型对象：如果模型对象正在使用，那么会放在值栈中 Action 的上面。如果某个 Action 应用了模型驱动，当 Action 被请求时，拦截器会自动从此 Action 中获得模型对象，并将所获得的对象放置在值栈中对应 Action 对象的上面。

（3）Action 对象：当每一个 Action 请求到来的时候，容器会创建一个此 Action 的对象并存入值栈，该对象携带所有与 Action 执行过程有关的信息。

（4）固定名称的对象：主要包括 Servlet 作用范围内相关的对象信息，这些对象包括#application、#session、#request、#attr 和#parameters。

访问值栈可以有很多方法，其中最常用的一种就是使用 JSP、Velocity 或者 Freemarker 提供的标签。还有就是使用 HTML 标签访问值栈中对象的属性；结合表达式使用控制标签（如 if、elseif 和 iterator）；使用 data 标签（set 和 push）来控制值栈本身。

值栈中对象的存储顺序如图 7.6 所示。

当在值栈中查找某个值时，会按照从上到下的顺序依次遍历每一个对象。如果在前一个对象中没找到需要的信息，则继续查找下一个对象，直到找到为止。使用值栈的优点是只需知道查找信息的“name”标识，即可在值栈中进行查找。

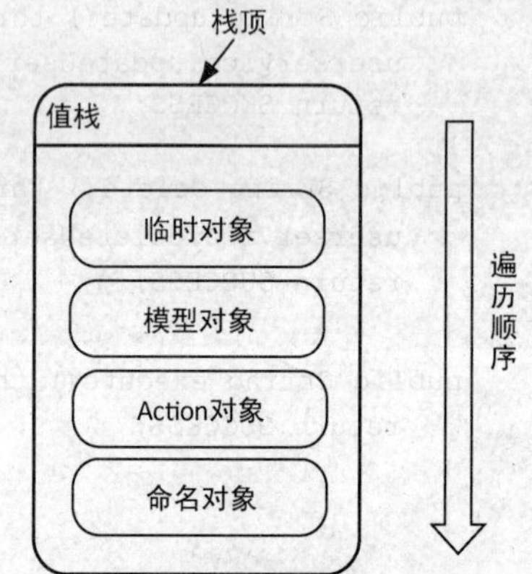

图 7.6　对象在值栈中的顺序

2. OGNL 表达式

Struts2 框架使用 OGNL 作为默认的表达式语言，Struts2 中很多地方都要用到 OGNL 表达式，如 Struts2 的标签、Struts2 的校验文件等。

OGNL 中有一个上下文（Context）概念，在 Struts2 中上下文（Context）的实现为 ActionContext，当 Struts2 接受一个请求时，会迅速创建 ActionContext、ValueStack（值栈）、Action。然后把 Action 存放进值栈中，所以 Action 的实例变量可以被 OGNL 访问。上下文中的对象包括 request、application、session、attr、parameters。访问上下文（Context）中的对象需要使用#符号标注命名空间，如#application、#session。

application 对象：访问当前应用程序中所有 ServletConfig 属性，例如#application.userName 或者#application['userName']，相当于调用 ServletContext 的 getAttribute("username")。

session 对象：访问当前 Session 中所有 HttpSession 属性，例如#session.userName 或者#session['userName']，相当于调用 session.getAttribute("userName")。

request 对象：访问当前请求中所有 HttpServletRequest 属性，例如#request.userName 或者#request['userName']，相当于调用 request.getAttribute("userName")。

parameters 对象：访问当前请求中所有 HttpServletRequest 参数，例如#parameters.userName 或者#parameters['userName']，相当于调用 request.getParameter("username")。

attr 对象：用于按 page->request->session->application 顺序访问其属性。

另外 OGNL 会设定一个根对象（root 对象），在 Struts2 中根对象就是 ValueStack（值栈）。如果要访问根对象（即 ValueStack）中对象的属性，则可以省略#命名空间，直接访问该对象的属性即可。

根对象 ValueStack 的实现类为 OgnlValueStack，该对象不是我们想像的只存放单个值，而是存放一组对象。在 OgnlValueStack 类里有一个 List 类型的 root 变量，用来存放一组对象。在 root 变量中处于第一位的对象叫栈顶对象，搜索顺序是从栈顶对象开始寻找。如果栈顶对象不存在该属性，就会从第二个对象寻找，如果没有找到就从第三个对象寻找，依次往下访问，直到找到为止。

OGNL 中同样提供了对集合元素的访问，通常情况下可以使用如下代码创建 List 对象：

```
<s:set name="list" value="{'zhangming','xiaoi','liming'}" />
<s:iterator value="#list" id="n">
   <s:property value="n"/><br>
</s:iterator>
```

OGNL 对 List 的访问与 Java 代码有所不同，如表 7.4 所示。

表 7.4　OGNL 与 Java 代码对 List 访问的比较

OGNL 访问的方式	Java 代码访问的方式
list[i]	list.get(i)
list.size	list.size()
list.isEmpty	list.isEmpty()

另外一种集合对象是 Map，可以使用下面的方式创建 Map：

```
<s:set name="foobar" value="#{'foo1':'bar1', 'foo2':'bar2'}" />
<s:iterator value="#foobar" >
   <s:property value="key"/>=<s:property value="value"/><br>
</s:iterator>
```

OGNL 创建 Map 是以“#”开头的，括号里的内容采用键值对的形式。在这里，set 标签用于将某个值放入指定范围。scope：指定变量被放置的范围，该属性可以接受 application、session、request、page 或 action。如果没有设置该属性，则默认放置在 OGNL Context 中（访问方式：#对象）。value：赋给变量的值，如果没有设置该属性，则将 ValueStack 栈顶的值赋给变量。OGNL 与 Java 代码对 Map 访问的比较如表 7.5 所示。

表 7.5　OGNL 与 Java 代码对 Map 访问的比较

OGNL 访问的方式	Java 代码访问的方式
map[‘one’]	map.get(“one”)
map.size	map.size()
map.isEmpty	map.isEmpty()

7.3.5　结果与视图

1. 配置 result 属性

Struts2 的 Action 处理完用户请求后返回一个普通字符串，整个普通字符串就是一个逻辑视图名。Struts2 通过在“struts.xml”文件中配置逻辑视图名和物理视图资源之间的映射关系，利用 Action 返回的字符串寻找到对应的物理视图，并呈现给用户。具体到“struts.xml”配置文件中，是通过设定 result 属性来实现的，result 按照作用范围可以分为局部 result 和全局 result。局部 result 包含在 Action 定义中，result 属性是一个 Action 的子元素，其作用范围只在本 Action 中；全局 result 使用<global-results…/>来定义，其作用范围是所有的 Action。下面是 result 的示例：

```
<!--定义全局返回类型 -->
<global-results>
    <result name="global-result">/welcome.jsp</result>
```

```
</global-results>
<action name="Action 名称" class="Action 类路径">
    <!--定义局部返回类型 -->
  <result name="逻辑视图名称" type="结果类型"</result>
  <result name="success">/hello.jsp</result>
    <param name="参数名称">参数值</param>
</action>
```

通过上面的示例，可以知道 result 元素包含 name 和 type 两个属性，其中 name 属性指定了逻辑视图的名称，type 属性指定了结果类型；param 元素的作用是为返回结果设置的参数。配置 result 元素时，name 属性的默认值为“success”，type 属性的默认值是“dispatcher”。通过上面的配置，若程序执行成功后则返回结果 success，系统会在“struts.xml”中寻找“name="success"”的 result 元素，找到后则页面跳转至指定的视图资源“hello.jsp”。

2. result 类型

type 属性指定了结果类型，默认情况下是 JSP 文件，除此以外，Struts2 还支持多种结果类型。Struts2 框架的内建的 result 类型，都是在 struts-default.xml 中定义的，系统会自动加载该 result 类型。struts-default.xml 中对 result 类型的定义如下：

```
<result-types>
  <result-type name="chain"
    class="com.opensymphony.xwork2.ActionChainResult"/>
  <result-type name="dispatcher"
    class="org.apache.struts2.dispatcher.ServletDispatcherResult"
    default="true"/>
  <result-type name="freemarker"
    class="org.apache.struts2.views.freemarker.FreemarkerResult"/>
  <result-type name="httpheader"
    class="org.apache.struts2.dispatcher.HttpHeaderResult"/>
  <result-type name="redirect"
    class="org.apache.struts2.dispatcher.ServletRedirectResult"/>
  <result-type name="redirectAction"
    class="org.apache.struts2.dispatcher.ServletActionRedirectResult"/>
  <result-type name="stream"
    class="org.apache.struts2.dispatcher.StreamResult"/>
  <result-type name="velocity"
    class="org.apache.struts2.dispatcher.VelocityResult"/>
  <result-type name="xslt"
    class="org.apache.struts2.views.xslt.XSLTResult"/>
  <result-type name="plainText"
    class="org.apache.struts2.dispatcher.PlainTextResult" />
  <result-type name="redirect-action"
    class="org.apache.struts2.dispatcher.ServletActionRedirectResult"/>
  <result-type name="plaintext"
    class="org.apache.struts2.dispatcher.PlainTextResult" />
</result-types>
```

Struts2 框架默认支持的 result 类型的名称和描述如下所述。

（1）chain：Action 链式处理的 result 类型。

（2）dispatcher：用来转向页面，通常处理 JSP，这是默认的结果类型。

（3）freemarker：用于整合 FreeMarker 的 result 类型。

（4）httpheader：用于处理特殊 HTTP 行为的 result 类型。

（5）redirect：用于重定向到一个 URL 的 result 类型。

（6）redirect-action：用于重定向到其他 Action 的 result 类型。

（7）stream：用于向浏览器返回一个 Inputstream，通常用来处理文件下载。

（8）velocity：用于返回结果对应视图 Velocity 模板。

（9）xslt：用于处理 XML/XSLT 的 result 类型。

（10）plaintext：用于显示页面的源代码的 result 类型。

Struts2 中 result 常见的几种转发类型仅有以下 4 种：dispatcher(默认)即内部请求转发，redirect 重定向，Action 链 chain 和 stream 文件下载。下面对这 4 种情况分别进行介绍。

（1）dispatcher：这种方式是 Struts2 中默认的转发类型，即内部请求转发，类似于 forward 的方式，而利用这种方式是将请求转发到一个页面，不可以转发到一个 Action。在配置结果类型为 dispatcher 时，需要注意以下几个方面。

- 请求转发只能将请求转发至同一个 Web 应用。
- 利用请求转发浏览器的地址栏不会发生变化。
- 利用请求转发调用者与被调用者之间共享相同的 Request 对象和 Response 对象，它们属于同一个访问的请求和响应。

（2）redirect

redirect 的配置代码片段如下：

```
<action?name="redirect">
<result?type="redirect">/login.jsp</result>
    </action>
```

这时候，用户在浏览器中访问该 Action，如输入 URL：http://localhost:8080/hello/redirect.action，当用户开始访问时浏览器的 URL 地址会变成：http://localhost:8080/hello/login.jsp。

上述代码片段是重定向到一个 JSP 页面。另外一种情况是 redirectAction，重定向到另一个 Action，这种配置往往在下面的情况下需要用到。例如，当管理员添加完一个用户后，系统自动跳转到用户列表的界面。配置代码示例如下：

```
<action?name="redirectAction">
<result?type="redirectAction">helloAction</result>
</action>
```

redirect 和 redirectAction 的区别和共同点如下：redirect 是在处理完当前 Action 之后，重定向到另外一个实际的物理资源；redirectAction 也是重定向，但它重定向到的是另外一个 Action。redirect 和 redirectAction 重定向均不共享 Request，重定向后的页面中无法接收 Request 里定义的参数等信息。

上文中所讲 dispatcher 请求转发和重定向之间的内部机制有着很大的区别，主要区别如下所述。

- 请求转发只能将请求转发给同一个 Web 应用中的组件；而重定向可以重新定向到同一站点不同应用程序中的资源，甚至可以定向到绝对的 URL 路径。
- 重定向可以看见目标页面的 URL；转发只能看见第一次访问的页面 URL。
- 请求响应类型中，调用者和被调用者之间共享相同的 Request 对象和 Response 对象；重定向调用者和被调用者属于两个独立访问请求和响应过程。
- 请求响应执行跳转页面后，下面的代码就不会继续执行；重定向跳转后还是会执行跳转后面的语句，因此必须加上 return。

另外，使用重定向返回类型时需要注意以下几个方面。

- 重定向不仅可以指定到一个 Web 应用，还可以指定到任何 JSP 资源。

- 重定向的访问结束后，浏览器的地址栏中显示 URL 的变化。
- 重定向的调用者和被调用者使用各自的 Request 对象和 Response 对象，它们属于两个独立的访问请求和响应过程。

（3）Action 链 chain

当一个 Action 执行完成后需要直接跳转到另一个 Action，此时就需要用到 Action 链。通过前文的学习可以知道，将 type 设置为 redirectAction 也可以实现跳转到另一个 Action。而 chain 和 redirectAction 的区别与 dispatcher 和 redirect 的区别是一样的，同样是跳转到一个 Action 上，但 chain 是服务器跳转，而 redirectAction 是客户端跳转。服务器跳转的过程中，可以共享数据，这时后面的 Action 就可以接收前面 Action 中的属性信息进行二次处理；而客户端跳转则不能共享数据。

Action 链的实现原理如图 7.7 所示。

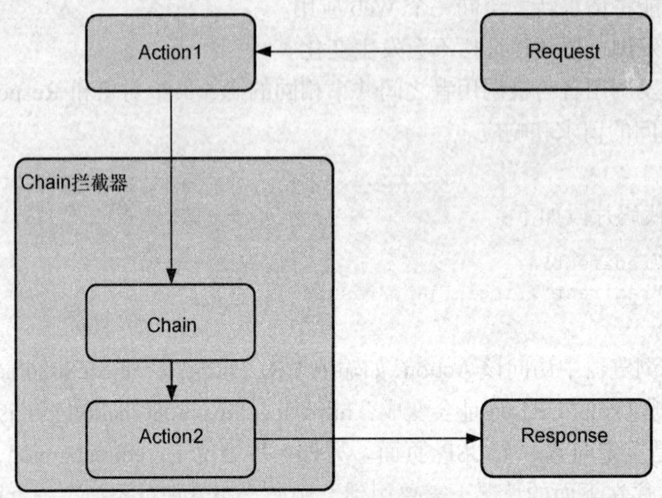

图 7.7　Action 链实现原理图

图 7.7 中，当 Action1 试图跳转到 Action2 中，则两个 Action 组成了一条 Action 链。当 Action1 的执行成功后，“chain”拦截器向 Action2 发出请求，Action2 开始执行。Action1 和 Action2 共享一个值栈，Action1 执行时信息会压入值栈，Action2 执行时若需要 Action1 中的数据，只需要从值栈中弹出即可。而 Action 链能起作用的必备条件是，Action2 中配置了“chain”拦截器。

（4）stream 文件下载

当 result 为 stream 类型时，Struts2 会自动根据配置好的参数下载文件。struts.xml 中配置相关信息的代码如下：

```
<struts>
<package name="default" extends="struts-default">
  <action name="download" class="action.DownloadAction">
    <result type="stream">
      <param name="contentType">application/octet-stream</param>
      <param name="inputName">inputStream</param>
      <param name="contentDisposition">attachment;filename="${fileName}"</param>
      <param name="bufferSize">4096</param>
    </result>
  </action>
</package>
</struts>
```

其中主要使用的参数如下所述。

contentType：指定下载文件的文件类型，application/octet-stream 表示无限制。

inputName：表示流对象名，例如 inputStream，则对应 Action 中的 getInputStream 方法。

contentDisposition：使用经过转码的文件名作为下载文件名，默认格式是“attachment; filename ="${fileName}"”，调用该 Action 中的 getFileName 方法。

bufferSize：下载文件的缓冲大小。

3. 动态返回结果

有些时候，只有当 Action 执行完毕的时候我们才知道要返回哪个结果，这个时候我们可以在 Action 内部定义一个属性，这个属性用来存储 Action 执行完毕之后的 result 值，例如：

```
private String nextAction;
public String getNextAction() {
  return nextAction;
}
```

在 struts.xml 配置文件中，可以使用${nextAction}来引用到 Action 中的属性，通过${nextAction}表示的内容来动态地返回结果，例如：

```
<action name="fragment" class="FragmentAction">
<result name="next" type="redirect-action">${nextAction}</result>
</action>
```

上述 Action 的 execute 方法返回 next 的时候，还需要根据 nextAction 的属性来判断具体定位到哪个 Action。

7.4 Struts2 深入理解

Struts2 框架中提供各种各样的技术，具有强大的功能。其中，拦截器是该框架中最重要的一项技术，其他的技术或多或少都会依赖拦截器，另外本节还会介绍 Struts2 的验证框架。

7.4.1 拦截器

拦截器是动态拦截 Action 时调用的对象，Struts2 大多数核心功能是通过拦截器实现的，每个拦截器完成某项特定功能，在 Action 执行之前或之后调用执行拦截器。早期 MVC 框架将一些通用操作写在核心控制器中，致使框架灵活性不足，可扩展性降低；Struts2 将核心功能放到多个拦截器中实现，拦截器可自由选择和组合，增强了灵活性，有利于系统的松耦合。

Struts2 拦截器栈（Interceptor Stack）是将拦截器按一定的顺序联结成的一条链。在访问被拦截的方法或字段时，Struts2 拦截器链中的拦截器就会按其之前定义的顺序被调用。

Struts2 拦截器的实现原理相对简单，当请求 Struts2 的 Action 时，Struts2 会查找配置文件，并根据其配置实例化相对的拦截器对象，然后串成一个列表依次调用列表中的拦截器。在 Action 和 Result 执行之后，拦截器再一次执行（与先前调用相反的顺序），在此链式的执行过程中，任何一个拦截器都可以直接返回，从而终止余下的拦截器、Action 及 Result 的执行。以下是一个定义拦截器和拦截器栈的实例：

```
<package name="mypackage" extends="struts-default" namespace="/manage">
<interceptors>
```

```
        <!-- 定义拦截器 -->
        <interceptor name="拦截器名" class="拦截器实现类"/>
        <!-- 定义拦截器栈 -->
        <interceptor-stack name="拦截器栈名">
           <interceptor-ref name="拦截器一"/>
           <interceptor-ref name="拦截器二"/>
        </interceptor-stack>
        </interceptors>
           ......
</package>
```

1. struts2 自带拦截器

在 Struts2 核心包中的“struts-default.xml”文件中定义了许多 Struts2 框架内置的多种功能的拦截器，下面将给出一些典型的 Struts2 自带拦截器。

（1）Params 拦截器：Params 拦截器提供了框架必不可少的功能，解析 HTTP 请求参数将其传送给 Action，设置成 Action 对应的属性值。

（2）staticParams 拦截器：在定义在 xml 配置文件中通过<action>元素的子元素<param>中的参数传入对应的 Action 的属性中。配置文件代码如下所示：

```
<action name="example" class="demo,ExampleAction">
   <param name="exampleField">Example</param>
   <result>success.jsp</result>
</action>
```

（3）servletConfig 拦截器：将源于 Servlet API 的各种对象注入到 Action，提供对 HttpServletRequest 和 HttpServletResponse 的访问机制。

（4）fileUpload 拦截器：对文件上传提供支持。

（5）exception 拦截器：捕获异常，并且将异常映射到用户自定义的错误页面。

（6）validation 拦截器：执行定义在 xxAction-validaiton.xml 中的校验器，调用验证框架进行数据验证。

（7）workflow 拦截器：提供当数据验证错误时终止执行流程的功能，即调用 Action 类的 validate()，校验失败返回 input 视图。

在很多情况下，一个 Action 需要使用多个拦截器，如果单独为每个 Action 配置这些拦截器，无论管理还是维护都会变得困难，这时可以使用拦截器栈。实际上拦截器栈也可以看作是一个拦截器，只不过是若干拦截器的集合。同样，“struts-default.xml”文件中定义了这些拦截器栈，这里并不一一列举。在这些拦截器栈中，defaultStack 拦截器栈被指定为默认拦截器，只要在定义包的过程中继承 struts-default 包，那么 defaultStack 将是默认的拦截器。

2. 配置拦截器

使用拦截器需要两个步骤：①通过<interceptor/>元素来定义拦截器；②通过<interceptor-ref/>元素使用拦截器。拦截器的 struts.xml 配置文件如下所示：

```
        <package name="mypackage" extends="struts-default" namespace="/manage">
            <interceptors>
                <interceptor name="interceptorName" class="interceptorClass" />   //定义拦
截器
                <interceptor-stack name="interceptorStackName">                  //定义拦
截器栈
```

```
            <interceptor-ref name="interceptorName|interceptorStackName1" />
          //指定引用的拦截器
        </interceptor-stack>
    </interceptors>
    <default-interceptor-ref name="interceptorName|interceptorStackName" />
    //定义默认的拦截器引用
    <action name="actionName" class="actionClass">
        <interceptor-ref name="interceptorName|interceptorStackName" />
         //为 Action 指定拦截器引用
         <!--省略其他配置-->
    </action>
</package>
```

3. 自定义拦截器

作为“框架（framework）”，可扩展性是不可或缺的。因此虽然 Struts2 提供如此丰富的拦截器实现需求，但是当内置拦截器不能满足需求时，可以允许自定义拦截器。创建拦截器的方式有两种：实现 Interceptor 接口和继承 AbstractInterceptor 类。

（1）实现 Interceptor 接口，该接口提供了 3 种方法，标准代码如下：

```
public abstract interface Interceptor extends Serializable {
    public abstract void destroy();
public abstract void init();
   public abstract String intercept(ActionInvocation invocation) throws Exception;
}
```

void init()：在该拦截器被初始化之后和执行拦截之前，系统回调该方法。对于每个拦截器而言，此方法只执行一次。

void destroy()：该方法跟 init()方法对应。在拦截器实例被销毁之前，系统将回调该方法，销毁 init()方法加载的资源。

String intercept(ActionInvocation invocation) throws Exception：该方法是用户需要实现的拦截动作，其参数为一个 ActionInvocation 对象，包含了被拦截的 Action 的引用。该方法会返回一个字符串作为逻辑视图，由系统根据字符串寻找到对应的物理视图资源进行跳转。

下面是一个拦截器的实例，用于在 Action 类调用前后输出提示信息，示例如下：

```
public class MyInterceptor implements Interceptor{
    public void destroy(){
    }
    public void init(){
    }
    public String intercept(ActionInvocation invocation) throws Exception{
        System.out.println("在 Action 执行前的拦截信息");   //Action 前置拦截
        String result = invocation.invoke();              //获得对应逻辑视图的字符串
        System.out.println("在 Action 执行后的拦截信息");
        return result;
    }
}
```

（2）继承 AbstractInterceptor 类：相比实现 Interceptor，继承 AbstractInterceptor 类的拦截器提供了 init()和 destroy()方法的空实现，只需要实现 interceptor 方法即可，代码如下：

```
public class MyInterceptor implements Interceptor{
    public String intercept(ActionInvocation invocation) throws Exception{
```

```
        System.out.println("在 Action 执行前的拦截信息");   //Action 前置拦截
        String result = invocation.invoke();             //获得对应逻辑视图的字符串
        System.out.println("在 Action 执行后的拦截信息");
        return result;
    }
```

7.4.2 Struts2 验证框架

输入校验是指在数据提交给程序处理之前，检查数据信息的合法性。输入校验可以保证数据的有效性和 Web 应用的安全性，是 Web 开发中一项非常重要的组成部分。输入校验分成客户端校验和服务端校验，它们缺一不可，都是用于防止非法的用户输入。客户端核验可以降低服务器负载，服务器端核验是请求数据进入系统的最后屏障。

客户端校验一般使用 JavaScript 脚本进行交互，代码支持正则表达式。在客户端可以进行许多校验功能防止非法的输入，例如，校验输入的文本是否为空，校验输入的文本格式是否正确，验证邮箱地址是否格式正确等。客户端校验可以快速地提示用户输入错误以便提高响应速度，更可以降低服务器的负担。

服务器校验也可以使用正则表达式，正则表达式可以匹配输入，再进行数据转换，减少抛出异常，提高程序效率。服务端校验的例子有：校验注册的用户名是否存在，校验登录的密码是否正确等。服务器校验的优点在于其安全性更高，另外在服务器端做的安全校验可以限制错误登录的次数、可以使用日志记录等；服务器校验也存在缺点，其执行效率通常比较低，如果用户响应时间过长，或发送了大量的校验请求产生拥塞时，用户等待时间将会变长。

1. 通过编码方式校验

在 Struts2 中可以通过编码方式实现校验，即在 Action 类中手动创建校验数据的代码。此处可以通过继承 ActionSupport 类，来完成 Action 的开发。其 execute()方法负责具体的业务逻辑，默认返回 SUCCESS，同时 ActionSupport 类还增加了对验证的支持。

若将大量的校验代码放在 execute()方法中，则会降低程序的可读性。一般方法是重写 validate()方法，从而实现校验逻辑和业务逻辑的分离。如果出现错误，则使用 addFieldError()和 addActionError()抛出错误信息，同时在 JSP 页面获取错误信息。示例代码如下：

```
public class LoginAction extends ActionSupport {
    public void validate() {
        if(getName().length()==0) {
        addFieldError("name","用户名不能为空");
        }
        else if (!Pattern.matches("\\w(4,15)", name)); {
        addFieldError("name","用户名必须是字母或数字且长度为 4~15 之间");
        }
        if(getPassword().length()==0) {
        addFieldError("password","密码不能为空");
        }
        ……省略其他验证
    }
    public String execute() throws Exception{
        return SUCCESS;
    }
}
```

2. 应用输入框架校验

通过在 Action 中添加 validate()方法，并手动编写代码可以实现校验，但这种方式存在缺点：当验证规则比较复杂时，需要编写烦琐的代码，且代码不可复用，维护相对困难。

鉴于数据验证的重要性和重复性，Struts2 中内置了一个验证框架，将常用的验证规则进行编码，使用框架时用户无需进行编码，只要在外部配置文件中指定某个字段需要进行的验证类型，并提供出错信息即可，从而提高开发效率。

Struts2 提供的常用校验器如表 7.6 所示。

表 7.6　校验器

名　称	可 选 参 数	描　　述	类　型
required	—	检查字段是否为空	字段
requiredstring	trim	检查字段是否为字符串且是否为空	字段
int	min，max	检查字段是否为整数，且是否在[min，max 之内]	字段
double	min，max	检查字段是否为双精度浮点数且在[min，max]之内	字段
date	min，max	检查字段是否为日期格式且在[min，max]之内	字段
expression	—	对指定 OGNL 表达式求值	非字段
fieldexpression	—	对指定 OGNL 表达式求值	非字段
email	—	检查字段是否为 E-mail 格式	字段
url	—	检查字段是否为 URL 格式	字段
visitor	context，appendPrefix	引用指定对象各属性对应的检验规则	字段
conversion	—	检查字段是否发生类型转换错误	字段
stringlength	trim，minLength，maxLength	检查字符串长度是否在指定范围内	字段
regex	—	检查字段是否匹配指定正则表达式	字段

下面举几个校验器的常用实例。

（1）requiredstring 校验器：判断用户名是否为空。

```
<validators type="requiredstring">
    <param name="trim">true</param>
    <message>用户名不能为空</message>
</validator>
```

上述代码中，<validators>元素用于声明校验器，其属性 type 值为校验器的名称；<param>元素用于指定要进行校验的字段，其属性 name 值指明校验器的类型；<message>指定校验失败时提示的信息。

（2）stringlength 校验器：判断用户名长度在 6～10 之内。

```
<field name="user.name">
     <field-validator type="stringlength">
     <param name="maxLength">10</param>
     <param name="minLength">6</param>
     <message>用户名长度须在${minLength}和${maxLength}之间</message>
</field-validator>
```

上述代码中<field>元素用于声明校验器，其允许在内部声明多个字段型或非字段型校验器，其 name 属性为要进行校验的字段名称。<field>元素通过<field-validator>元素引入需要的校验器，其 type 值为校验器的名称，同样<message>指定校验失败时提示的信息。

7.5 Struts2 编程实例

前面已经介绍了 Struts2 的框架基础和内部的工作流程，下面将以一个简单的应用程序为例，介绍该框架如何拦截客户请求，如何调用业务控制器处理请求等内容。该应用是一个简单的登录模块设计，允许用户输入用户名和密码，如果符合要求，则进入登录页面，如果不符合要求，则会进入一个提示页面。

7.5.1 Struts2 安装配置

Struts2 的开发依赖于 Java 环境，按照第 1 章的内容，在机器上安装 JDK、Tomcat 和 MyEclipse，本节介绍如何下载和安装 Struts2 框架。

首先，从 http://struts.apache.org/download.cgi 下载最新的 Struts2 版本，如图 7.8 所示。

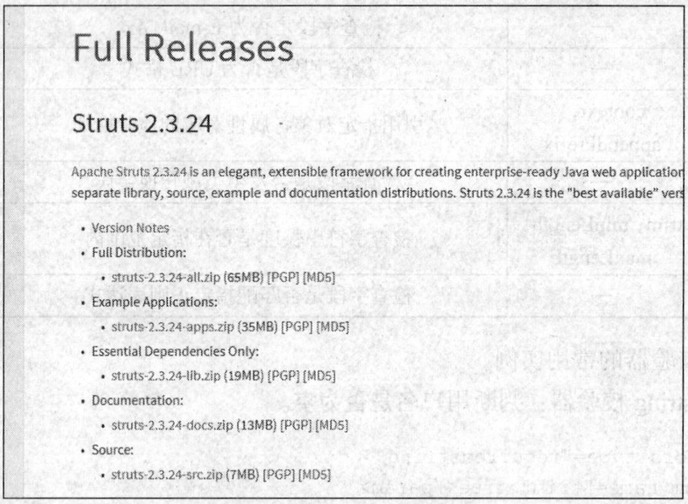

Full Releases

Struts 2.3.24

Apache Struts 2.3.24 is an elegant, extensible framework for creating enterprise-ready Java web applicatio
separate library, source, example and documentation distributions. Struts 2.3.24 is the "best available" vers

- Version Notes
- Full Distribution:
  - struts-2.3.24-all.zip (65MB) [PGP] [MD5]
- Example Applications:
  - struts-2.3.24-apps.zip (35MB) [PGP] [MD5]
- Essential Dependencies Only:
  - struts-2.3.24-lib.zip (19MB) [PGP] [MD5]
- Documentation:
  - struts-2.3.24-docs.zip (13MB) [PGP] [MD5]
- Source:
  - struts-2.3.24-src.zip (7MB) [PGP] [MD5]

图 7.8　Struts2 下载界面

在该界面有如下选项。

Full Distribution：Struts2 的完整版，通常建议下载该选项。

Example Applications：Struts2 的示例应用，这些示例应用对于学习 Struts2 有很大的帮助，下载 Struts2 的完整版时，已经包含了该选项下的全部应用。

Blank Application only：下载 Struts2 的空示例应用，该示例已经包含在 Example Applications 选项下。

Essential Dependencies：下载 Struts2 的核心库。

Documentation：仅仅下载 Struts2 的相关文档，包含 Struts2 的使用文档、参考手册和 API 文档等。

Source：下载 Struts2 的全部源代码。

Alternative Java 4 JARs：下载可选的 JDK 1.4 的支持 JAR。

这里通常建议读者下载第一个选项：Full Distribution。将下载到的 zip 文件解压缩，该文件就是一个典型的 Web 结构，该文件夹应包含如下文件结构。

apps：包含基于 Struts2 的示例应用，这些示例应用对于学习者是非常有用的资料。

docs：包含 Struts2 的相关文档，包括 Struts2 的快速入门、Struts2 的文档，以及 API 文档等内容。

j4：包含了让 Struts2 支持 JDK 1.4 的 JAR 文件。

lib：包含 Struts2 框架的核心类库，以及 Struts2 的第三方插件类库。

src：包含 Struts2 框架的全部源代码。

在开发 Web 应用时，将 lib 文件夹下的 Struts2-core-2.3.24.jar、xwork-2.0.1.jar 和 ognl-2.6.11.jar 等必需类库复制到 Web 应用的 WEB-INF/lib 路径下。当然，如果你的 Web 应用需要使用 Struts2 的更多特性，则需要将更多的 JAR 文件复制到 Web 应用的 WEB-INF/lib 路径下。但大部分时候并不需要利用到 Struts2 的全部特性，因此没有必要一次将该 lib 路径下 JAR 文件全部复制到 Web 应用的 WEB-INF/lib 路径下。

最后编辑 Web 应用的 web.xml 配置文件，配置 Struts2 的核心 Filter，部分代码如下：

```
<filter>                                         //定义 Struts 2 的 FilterDispatcher 的 Filter
<filter-name>struts2</filter-name>               //定义核心 Filter 的名字
<filter-class>org.apache.Struts2.dispatcher.FilterDispatcher
//定义核心 Filter 的实现类
</ filter-class>
</filter>
<!-- FilterDispatcher -->                        //初始化 Struts 2 并且处理所有的 Web 请求
<filter-mapping>
<filter-name>Struts2</filter-name>
<url-pattern>/*</url-pattern>
</filter-mapping>
```

经过上述 3 个步骤，至此已经可以在一个 Web 应用中使用 Struts2 的基本功能了，后文将具体介绍 Struts2 在 Web 应用中的编程实例。

7.5.2　创建 Struts2 的 Web 应用

打开前文章节中介绍的 MyEclipse，新建一个 Web Project，可自行定义名称，添加 xml 文件。工程建立好后，在 WEB-INF 路径下的 lib 库中添加 Struts2 的框架类库，并在 WEB-INF 路径下添加 struts.xml 配置文件。

1. 实现请求页面

本应用使用传统的 JSP 视图技术：当用户需要登录本系统时，填写表单提交页面，这个表单提交页面涉及两个表单域，即用户名和密码。该页面是一个静态的 HTML 页面，使用了 Struts2 的重要组件标签库，并且使用了 Action 属性，当表单提交给 login.action 时，Struts2 的 FilterDispatcher 将用户请求转发到对应的 Struts2 的 Action。例 7-8 是用户请求登录的 JSP 页面。

例 7-8　用户请求登录的 JSP 页面。

```
<%@ page language="java" contentType="text/html; charset=GBK"%>
<%@taglib prefix="s" uri="/struts-tags"%>
<html>
<head>
<title>登录页面</title>
```

```
</head>
<body>
<!-- 使用 form 标签生成表单元素 -->
<s:form action="Login">
    <!-- 生成一个用户名文本输入框 -->
    <s:textfield name="username" label="用户名"/>
    <!-- 生成一个密码文本输入框 -->
    <s:textfield name="password" label="密码"/>
    <!-- 生成一个提交按钮->
    <s:submit value="登录"/>
</s:form>
</body>
</html>
```

2. 实现控制器

MVC 的框架核心就是控制器，当用户通过上文的页面提交用户请求时，该请求需要交给控制器处理，控制器根据处理结果决定将哪个页面返回给用户。例 7-9 是处理用户请求的 Action 代码。

例 7-9　处理用户请求的 Action 代码。

```
public class LoginAction implements Action    //Struts2 的 Action 类是普通的 Java 类
{
    private String username;
    private String password;                     //Action 内封装用户请求的两个参数
    public String getUsername()                  //username 属性对应的 getter 方法
    {
        return username;
    }
    public void setUsername(String username)        //username 属性对应的 setter 方法
    {
        this.username = username;
    }
    public String getPassword()                  //password 属性对应的 getter 方法
        {
        return password;
     }
     public void setPassword(String password)     //password 属性对应的 setter 方法
     {
        this.password = password;
     }
     public String execute() throws Exception       //处理用户请求的 execute 方法
     {
        if (getUsername().equals("scott") && getPassword().equals("tiger") )
        {
            ActionContext.getContext().getSession().put("user" , getUsername())
            return SUCCESS;
        }
       else
       {
          return ERROR;
       }
    }
```

```
    public void validate()
    {
        if (getUsername() == null || getUsername().trim().equals(""))
        //如果用户名为空或为空字段，添加表单校验错误
        {
            addFieldError("username", "user.required");
        }
        if (getPassword() == null || getPassword().trim().equals(""))
        //当密码为空或空字符串时，添加表单校验错误
        {
            addFieldError("password", "pass.required");
        }
     }
}
```

上面的代码重写了 validate()方法，该方法会在执行系统的 execute()方法之前执行，如果执行该方法后，Action 类的 fieldErrors 中已经包含了数据校验错误，则请求被转发到 input 逻辑视图处。当然也可以使用 Struts2 的校验框架，需要增加一个校验配置文件，请读者课后尝试使用 Struts2 的校验框架完成对用户名和密码的验证。

代码中定义了 Action，而当 Action 处理完用户登录时，还需要跟踪用户登录的状态信息——通常当一个用户登录后，需要将用户的用户名添加为 Session 状态信息。为了访问 HttpSession 实例，我们通过 ActionContext 的 getSession 方法访问 Web 应用中的 Session。此时拦截器则派上用场，由于 getSession 返回的不是 HttpSession 对象，拦截器负责该 Session 和 HttpSession 的转换。

下面所要做的就是在“struts.xml”中配置 Action，配置文件中处理指定 Action 的实现类，还需要给出 Action 处理结果与资源之间的映射关系，下面是配置文件中的代码。

```
<?xml version="1.0" encoding="GBK"?>
<!DOCTYPE struts public
     "-//Apache Software Foundation//DTD Struts Configuration 2.0//
     "http://struts.apache.org/dtds/struts-2.0.dtd">
<!-- struts 是 Struts 2 配置文件的根元素-->
<struts>
   <package name="mypackage" extends="struts-default">
       <!-- 定义 login 的 Action，该 Action 的实现类为 LoginAction 类 -->
       <action name="Login" class=" LoginAction">
           <!-- 结果与资源的映射关系 -->
           <result name="input">/login.jsp</result>
           <result name="error">/error.jsp</result>
           <result name="success">/success.jsp</result>
       </action>
   </package>
</struts>
```

上面的映射文件定义了 name 为“Login”的 Action，该 Action 调用自身的 execute 方法处理用户请求，如果返回 SUCCESS，将转发到/success.jsp 页面，而如果返回 error 字符串，则请求转发到/error.jsp 页面。

3. 增加视图资源

下面的任务是增加两个视图页面，success.jsp 和 error.jsp，将这两个文件放在 Web 应用的根目录下，success.jsp 的代码如下所示：

```
<%@ page language="java" contentType="text/html; charset=GBK"%>
<html>
<head>
    <title>成功页面</title>
</head>
<body>
    欢迎，${sessionScope.user}，您已经登录！
</body>
</html>
```

error.jsp 页面也与 success 页面类似，请读者自行完成。

思考与习题

1. 填空题：MVC 是一种优秀的软件设计模式，它将应用分成______、______和______3 个层次，从而使同一个应用程序使用不同的表现形式。

2. 填空题：值栈中的对象包括：临时对象、______、______和______。

3. 填空题：OGNL 全称为______________，是一种功能强大的表达式语言。

4. 简答题：简单介绍 MVC 设计模式和 Struts2 工作流程。

5. 简答题：简单介绍 Struts2 框架默认支持的几种 result 类型中，dispatcher 请求转发和重定向之间内部机制的区别。

6. 简答题："struts.xml"文件中，<action>元素都有哪些属性？作用是什么？

7. 编程题：针对上一章思考与习题中编写的注册系统，添加验证模块。主要验证：用户名、邮箱、密码、手机号码等是否合法。

第8章 工业园区企业安全巡检系统

8.1 系统设计

8.1.1 开发背景和需求分析

1. 开发背景

现如今，工业园区的产业快速发展，经济发展与安全环境之间的关系难以协调，企业在追求经济效益最大化的同时，安全工作方面却存在着诸多问题。各地工业园区发生火灾或爆炸等情况屡见不鲜，这不仅给园区企业的经济效益带来损失，更给人身安全带来严重威胁。

随着社会的进步与发展，各行业对其管理工作的要求越来越规范化、科学化。各单位对定期定点进行维护、检测人员的责任心的要求也越来越高。针对存在多种安全隐患的工业园区而言，据不完全统计，园区内发生的灾难大部分可以通过安全巡检的方式防患于未然。因此，随着对安全的呼声越来越高，安全巡检工作的重要性也日益突出。

2. 需求分析

传统的工厂和企业安全巡检工作主要是通过安全员的定时巡查和上层领导的不定期抽查两种方式，安全人员负责记录安全情况，排查安全隐患。但是这种考核方式并不科学，在核实安全员是否每天准时到达现场方面，很难做到统计真实的情况。而这种考核方式容易让巡检员产生懈怠心理，很可能因为一个小的失误或操作，或者未到岗排查安全隐患，从而导致灾难性的后果。

传统的方式有以下缺点：

（1）人为因素多，管理成本高；

（2）无法实时监督巡检员的工作；

（3）安全巡检数据的信息化程度低；

（4）巡检数据不能实时进行分析处理，影响效率。

基于以上不合理因素的考虑，园区安全巡检平台系统通过手机端的 GPS 定位技术和无线网络可以将安全员的定位信息和巡检数据、状态过程信息实时发布至系统，这样可以杜绝巡检人员和被巡检对象无法科学、准确考核与监控的现象。同时，系统平台的科学化信息管理可以为巡检维护管理工作提供有效的手段。系统能够有效地提高各类巡检工作的规范化及科学化水平，有效地保障被巡检设施处于良好状态。

8.1.2　系统目标与功能结构

平台分为手机 APP 端和 Web 管理信息系统，手机端可以实时统计，Web 端管理园区内各个企业的信息，以及安全巡检记录的信息。以下我们所介绍的系统均为基于 Web 的管理信息系统。

1. 系统目标

根据系统需求分析和与客户的沟通，工业园区安全巡检系统需要达到以下目标。

（1）系统界面设计友好、简洁大方美观。

（2）在首页中提供用户登录功能。

（3）用户登录后，系统赋予不同用户角色不同的操作权限。企业信息模块中，每个企业的用户可以查看自己企业的信息，但不提供对其他企业信息查看的功能；企业信息包含企业名称、类型、电话、法人、成立时间、企业安全级别等。企业安全级别对应于不同企业类型所需设置的安全级别，不同的级别对应不同的监督力度，从而达到资源的最大化利用。

（4）监督管理模块中，超级管理员可以增删改查企业所在的位置、企业监督点位置信息、安检员上传的巡检记录；普通用户可以增删改查该企业的监督点相关信息。

（5）统计分析模块中，包含安检员的基本信息、出勤率、出勤时间等，并以图表形式展现。

（6）系统具有易维护性和易操作性。

2. 功能结构

系统分为前、后台两部分设计，前台主要面向访问用户，实现数据信息的展示、查询与添加功能，其中信息的显示包括列表显示、图表显示和详细内容的显示。后台主要负责管理前台的数据操作。

3. 开发环境

在开发系统时，需要具备以下开发环境。

（1）服务器端

操作系统：Windows 7 操作系统。

Web 服务器：Tomcat 7.0。

Web 框架：Spring、DWR、Mybatis。

Java 开发包：JDK 1.7。

数据库：MySQL。

浏览器：IE 7.0。

分辨率：最佳效果为 1024 像素 × 768 像素。

（2）客户端

浏览器：IE 7.0。

分辨率：最佳效果为 1024 像素 × 768 像素。

8.1.3　数据库设计

本系统是一个中小型的管理信息系统，考虑到开发成本、用户信息量及客户需求等问题，系统采用的是 MySQL 作为项目的数据库。MySQL 是流行的关系型数据库管理系统，在 Web 应用方面，MySQL 是最好的关系数据库管理系统（Relational Database Management System，RDMS）应用软件之一。MySQL 是一种关联数据库管理系统，关联数据库将数据保存在不同的表中，而不是将所有数据放在一个大仓库内，这样就增加了速度并提高了灵活性。MySQL 所使用的 SQL

语言是用于访问数据库的最常用标准化语言。

根据项目的需求，需要创建与实体对应的数据表，它们分别为数据表：企业类型表、企业安全级别表、企业信息表、企业巡检点表、企业用户表、角色表、出勤记录表，以及巡检安排表，其中用户和角色表之间相互关联。

图 8.1 ~ 图 8.6 给出了其中企业信息表、企业巡检点表、企业类型表、企业安全级别表、企业出勤记录表和企业用户表的表结构设计图。

企业信息表		
企业ID	int	<pk>
标识编号	varchar(100)	
企业名称	varchar(100)	
企业类型	varchar(100)	
安全级别	varchar(100)	
法人	varchar(100)	
成立时间	date	
经营范围	varchar(2000)	
企业规模	varchar(100)	
注册资本	varchar(100)	
注册地址	varchar(200)	
负责人	varchar(30)	
联系电话	varchar(30)	
有效标识	char(1)	
备注	varchar(2000)	
经度	varchar(30)	
纬度	varchar(30)	

图 8.1　企业信息表

企业巡检点表	
巡检点ID	int
巡检点名称	varchar(100)
所属企业	int
负责人	varchar(100)
联系电话	varchar(30)
安全规范	varchar(2000)
巡检说明	varchar(2000)
经度	varchar(30)
纬度	varchar(30)
误差	varchar(30)
上传频率	varchar(30)
设定时间	timestamp
有效标识	char(1)
是否存在隐患	char(1)

图 8.2　企业巡检点表

企业类型表		
企业类型ID	int	<pk>
类型编号	varchar(100)	<ak>
企业类型	varchar(100)	
有效标识	char(1)	

图 8.3　企业类型表

企业安全级别表		
企业类型ID	int	<pk>
类型编号	varchar(100)	<ak>
企业类型	varchar(100)	
有效标识	char(1)	

图 8.4　企业安全级别表

出勤记录表		
出勤ID	int	<pk>
巡检安排ID	int	
出勤时间	datetime	
出勤经度	varchar(10)	
出勤纬度	varchar(10)	
终端ID	varchar(30)	
企业ID	int	
安监员ID	int	
手机号	varchar(30)	
出勤状态	char(1)	
距离	varchar(30)	

图 8.5　出勤记录表

企业用户表		
用户ID	int	<pk>
用户姓名	varchar(30)	
用户账号	varchar(30)	
用户密码	varchar(30)	
所属企业	int	
手机号	varchar(30)	<ak>
岗位	varchar(30)	
有效标识	char(1)	
创建时间	timestamp	
备注	varchar(2000)	

图 8.6　企业用户表

8.1.4 系统预览图

图 8.7 和图 8.8 所示是本章介绍系统的登录页面效果图和首页效果图。

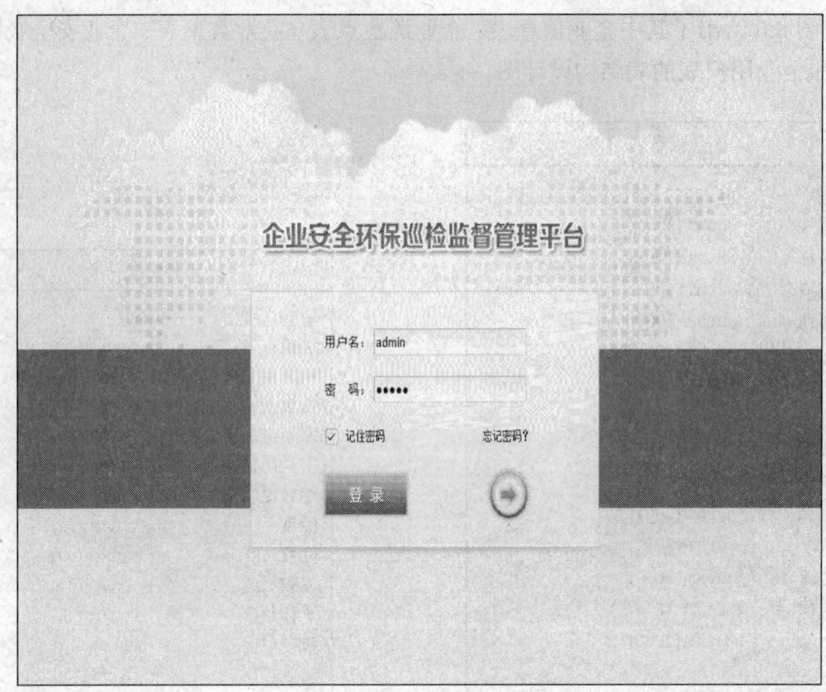

图 8.7 登录页面效果图

图 8.8 首页效果图

8.2　Spring 框架介绍

在本章的编程实例“工业园区企业安全巡检系统”实现中，用到了实用的 Web 框架 Spring、DWR、Mybatis 和相应的技术等，下面我们根据代码实例对这些框架和技术进行介绍，Mybatis 框架在第 9 章实例中介绍。

8.2.1　Spring 基础

Spring 是于 2003 年兴起的一个轻量级的 Java 开源开发框架，是由 Rod Johnson 在其著作中阐述的部分理念和原型衍生而来。该框架的主要优势之一就是其分层架构，分层架构允许使用者选择使用哪一个组件，同时为 J2EE 应用程序开发提供集成的框架。

Spring 的用途不仅限于服务器端的开发，从简单性、可测试性和松耦合的角度而言，任何 Java 应用都可以从 Spring 中受益。Spring 的核心是控制反转（IoC）和面向切面（AOP）。简单来说，Spring 是一个分层的 JavaSE/EEfull-stack（一站式）轻量级开源框架。

1. 框架特征

轻量：从大小与开销两方面而言，Spring 都是轻量的。完整的 Spring 框架可以在一个大小只有 1MB 多的 JAR 文件里发布，并且 Spring 所需的处理开销也是微不足道的。

控制反转：Spring 通过一种控制反转（IoC）的技术促进了低耦合。当应用了 IoC，一个对象依赖的其他对象会通过被动的方式传递进来，而不是这个对象自己创建或者查找依赖对象。

面向切面：Spring 提供了面向切面编程（AOP）的丰富支持，允许通过分离应用的业务逻辑与系统级服务进行内聚性的开发。应用对象只实现业务逻辑，并不负责其余关注点，如日志或事务支持。

容器：Spring 包含并管理应用对象的配置和生命周期，在这个意义上它是一种容器。用户可以基于一个可配置原型（prototype），配置每个 bean。用户的 bean 可以创建一个单独的实例，或者每次需要时都生成一个新的实例。

框架：Spring 可以将简单的组件配置、组合成为复杂的应用。在 Spring 中，应用对象在一个 XML 文件被声明。Spring 也提供了很多基础功能，如事务管理、持久化框架集成等，由用户自己进行应用逻辑的开发。

MVC：Spring 的作用不仅仅限于整合，Spring 框架中，客户端发送请求，服务器控制器（由 DispatcherServlet 实现的)完成请求的转发，控制器调用一个用于映射的类 HandlerMapping，该类用于将请求映射到对应的处理器来处理请求。Spring 框架中还提供一个视图组件 ViewResolver，该组件根据 Controller 返回的标示找到对应的视图，将响应 response 返回给用户。

2. Spring 处理请求的流程

Spring 框架也是一个基于请求驱动的 Web 框架，并且也使用了前端控制器模式来进行设计，再根据请求映射规则分发给相应的页面控制器（动作/处理器）进行处理。下面是 Spring 处理请求的流程。

（1）用户发送请求至前端控制器。

（2）前端控制器根据请求信息（如 URL）来决定选择哪一个页面控制器进行处理并把委托请求发送给该页面控制器。

（3）页面控制器接收到请求后，进行功能处理，首先需要收集和绑定请求参数到一个对象，这个对象在 Spring Web MVC 中叫命令对象，并进行验证，然后将命令对象委托给业务对象进行处理；处理完毕后返回一个 Model And View（模型数据和逻辑视图名）。

（4）前端控制器收回控制权，然后根据返回的逻辑视图名，选择相应的视图进行渲染，并把模型数据传入以便视图渲染。

（5）前端控制器再次收回控制权，将响应返回给用户。

8.2.2 Spring 骨骼架构

Spring 总共有十几个组件，但是真正实用的组件只有几个，图 8.9 所示是 Spring 框架的总体架构图。

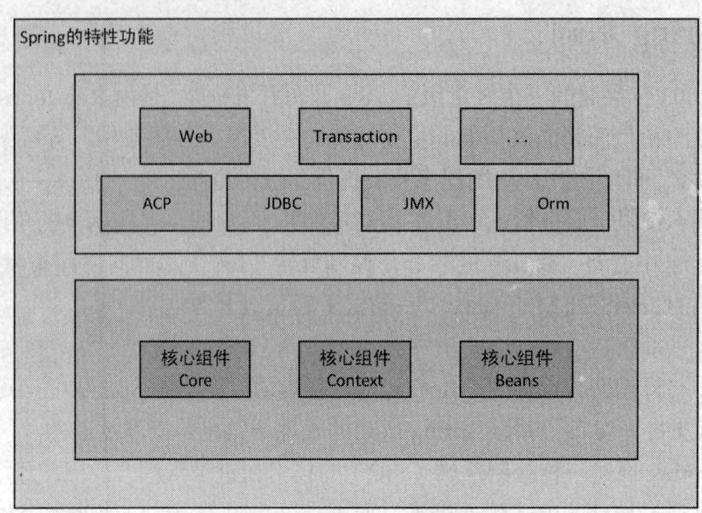

图 8.9 Spring 框架总体架构图

从图 8.9 中可以看出 Spring 框架中的核心组件只有 3 个：Core、Context 和 Beans。它们构建起了整个 Spring 的骨骼架构。没有它们就不可能有 AOP、Web 等上层的特性功能。下面也将主要从这 3 个组件入手分析 Spring。

Bean 在 Spring 中的作用很重要，Spring 中没有 Bean 也就没有 Spring 存在的意义。Spring 中关键的机制就是依赖注入机制，可以将对象之间的依赖关系转而用配置文件来管理，而这个注入关系在一个叫 Ioc 容器中管理，那 Ioc 容器存放着的正是被 Bean 包裹的对象。Spring 正是通过把对象包装在 Bean 中而达到对这些对象管理及一系列额外操作的目的。

Context 就是一个 Bean 关系的集合，这个关系集合即为上一段提到的 Ioc 容器。Context 主要的工作是为包装在 Bean 中对象的数据提供生存环境，具体来说，就是要发现每个 Bean 之间的关系，为它们建立这种关系并且要维护好这种关系。

Core 组件的作用是发现、建立和维护每个 Bean 之间的关系所需要的一系列的工具。

1. Bean 组件

Bean 组件在 Spring 的 org.springframework.beans 包下。这个包下的所有类主要解决了三件事：Bean 的定义、Bean 的创建，以及对 Bean 的解析。

Spring Bean 的创建是典型的工厂模式，它的顶级接口是 BeanFactory，BeanFactory 有 3 个子类：ListableBeanFactory、HierarchicalBeanFactory 和 AutowireCapableBeanFactory。它们最终的默认实现类是 DefaultListableBeanFactory，这 4 个接口共同定义了 Bean 的集合、Bean 之间的关系，以及 Bean 行为。

Bean 的定义主要有 BeanDefinition 描述，Bean 的定义就是完整地描述了在 Spring 的配置文件中定义的<bean/>节点中所有的信息，包括各种子节点。当 Spring 成功解析一个定义的<bean/>节点后，在 Spring 的内部，它就被转化成 BeanDefinition 对象，如图 8.10 所示。

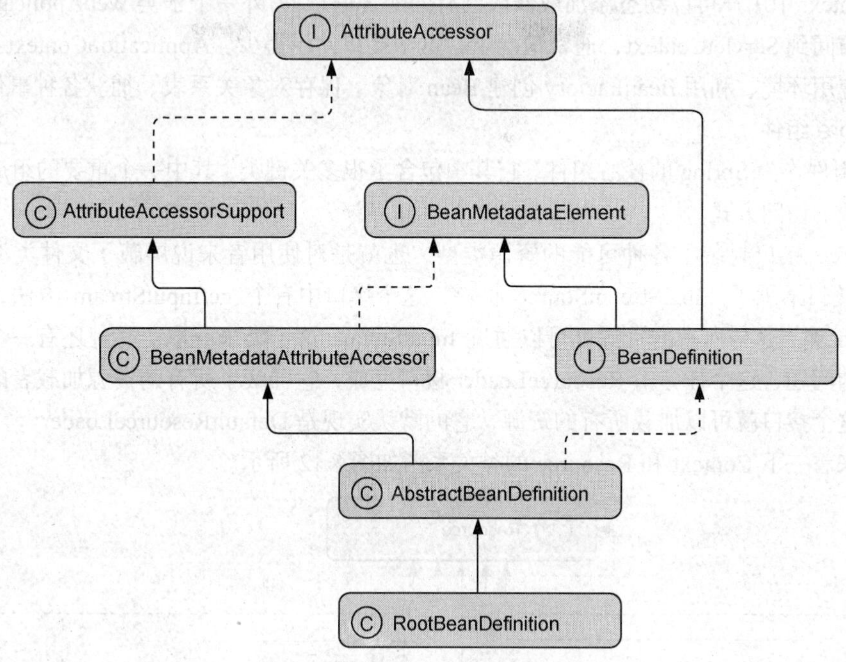

图 8.10　Bean 定义的类层次关系图

Bean 的解析过程非常复杂，功能被分得很细，因为这里需要被扩展的地方很多，必须保证有足够的灵活性，以应对可能的变化。Bean 的解析主要就是对 Spring 配置文件的解析。解析过程主要通过图 8.11 中的类完成。

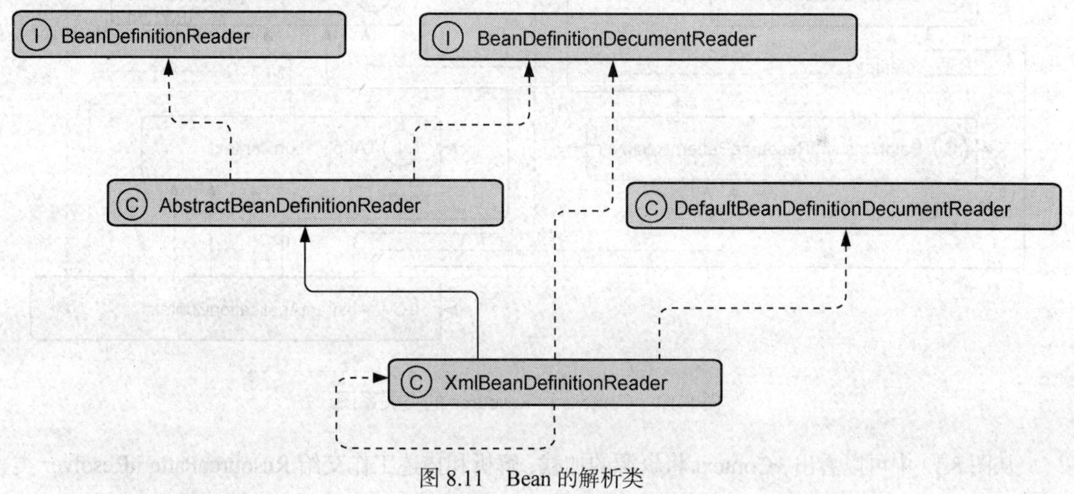

图 8.11　Bean 的解析类

2. Context 组件

Context 组件为 Spring 提供一个运行时的环境，用以保存各个对象的状态。ApplicationContext 是 Context 的顶级父类，ApplicationContext 继承了 BeanFactory，这说明 Spring 容器中运行的主体对象是 Bean。另外，ApplicationContext 继承了 ResourceLoader 接口，使 ApplicationContext 可以

访问到任何外部资源。

ApplicationContext 的子类 ConfigurableApplicationContext 表示该 Context 是可修改的，也就是在构建 Context 中用户可以动态添加或修改已有的配置信息；另外一个子类 WebApplicationContext 可以直接访问到 ServletContext，通常情况下，这个接口使用得少。ApplicationContext 主要功能：标识一个应用环境、利用 BeanFactory 创建 Bean 对象、保存对象关系表、捕获各种事件。

3. Core 组件

Core 组件作为 Spring 的核心组件，它其中包含了很多关键类，其中一个重要的组成部分就是定义了资源的访问方式。

Resource 接口封装了各种可能的资源类型，也就是对使用者来说屏蔽了文件类型的不同。Resource 接口继承了 InputStreamSource 接口，这个接口中有个 getInputStream 方法，返回的是 InputStream 类。这样所有的资源都可以通过 InputStream 这个类来获取。另外还有一个问题就是加载资源的问题，这个任务由 ResourceLoader 接口完成，它屏蔽了所有的资源加载者的差异，只需要实现这个接口就可以加载所有的资源，它的默认实现是 DefaultResourceLoader。

首先来看一下 Context 和 Resource 的类关系图如图 8.12 所示。

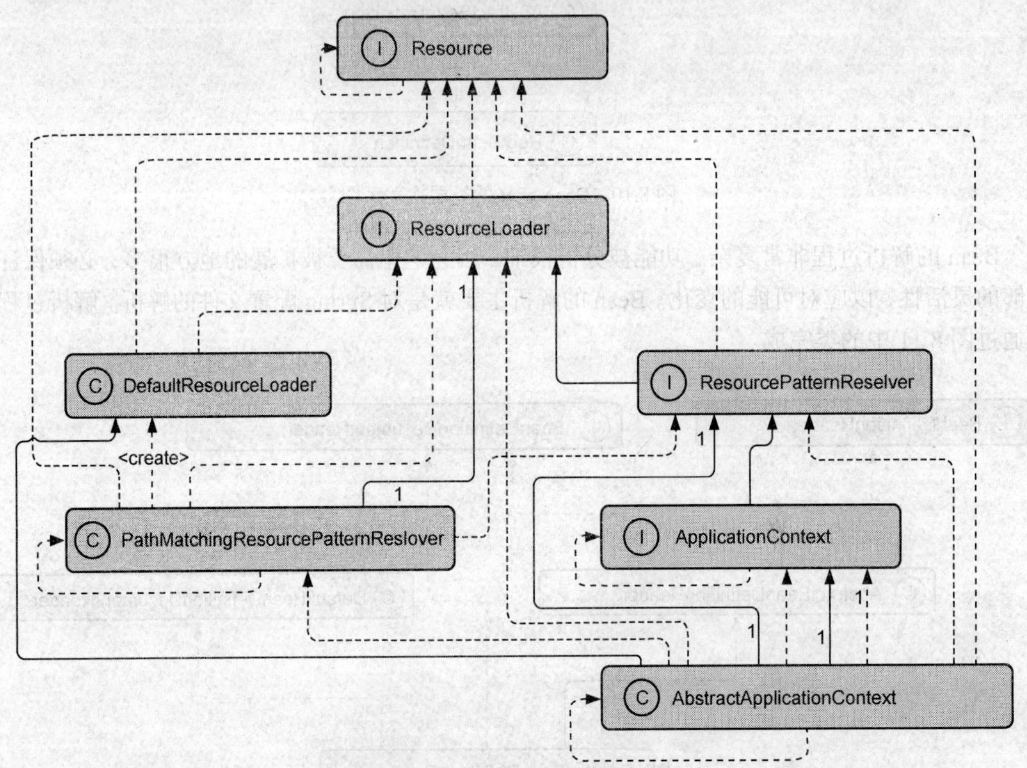

图 8.12　Context 和 Resource 的类关系图

从图 8.12 中可以看出，Context 将资源的加载、解析和描述工作交给 ResourcePatternResolver 类完成，ResourcePatternResolver 类把资源的加载、解析和资源的定义整合在一起便于其他组件使用。

Core 组件中还有很多类似的方式。

8.2.3 Bean 的装配

Spring 中要想使用容器管理 Bean，就必须在配置文件中规定好各个 Bean 的属性及依赖关系。通过配置此信息，容器才知道何时创建一个 Bean、何时注入一个 Bean，以及何时销毁一个 Bean。

1. Bean 的基本配置

一个最基本的 Bean 配置应该包含以下内容：Bean 的名称和 Bean 对应的基本类。例如下面代码所示，当在配置文件中用到指定的“XX”时，IoC 容器就会根据配置内容实例化相应的类，其基本格式如下所示：

```
<?xml version="1.0" encoding="UTF-8"?>
<beans xmlns="http://www.springframework.org/schema/beans"
   xmlns:xsi="http://www.w3.org/2001/XMLSchema-instance"
   xsi:schemaLocation="http://www.springframework.org/schema/beans
   http://www.springframework.org/schema/beans/spring-beans-3.0.xsd
<bean id="XX" classes="YY"/>
</beans>
```

Bean 的名称不可重复，每个 Bean 的 id 必须是唯一的。如果某个 Bean 需要使用命名规则以外的命名方式，可以使用“name”属性代替“id”属性。当使用“name”属性时，可以为 Bean 指定多个名称，中间用逗号分隔。

2. Bean 的属性

Bean 可以通过<property>元素为 Bean 组件添加属性，该属性可以是一个变量、一个集合，或者对其他 Bean 的一个引用。

为 Bean 添加一个变量的示例代码如下：

```
<bean id="XX" class="YY">
     <property name="locations">
     <value>/WEB-INF/jdbc.oracle.properties</value>
     </property>
</bean>
```

多数情况下，各个 Bean 都不是独立存在的，而是与其他的发生关联。当引用其他的 Bean 时，就需要用到<property>的<ref>元素。例如下面代码，在“sqlSessionFactory”中引用了“dataSource”和“configLocation”。

```
<bean id="sqlSessionFactory" class="org.mybatis.spring.SqlSessionFactoryBean">
     <property name="dataSource" ref="dataSource" />
     <property name="configLocation" value="classpath:mybatis/configuration.xml" />
</bean>
```

下面给出本章所介绍的工业园区安全巡检系统中，Spring 框架最为核心的配置文件“applicationContext.xml”中的配置代码，和“web.xml”中关于 Spring 框架的配置代码。

applicationContext.xml：

```
<?xml version="1.0" encoding="UTF-8"?>
<beans xmlns="http://www.springframework.org/schema/beans"
   xmlns:xsi="http://www.w3.org/2001/XMLSchema-instance"
   xmlns:context="http://www.springframework.org/schema/context"
   xmlns:aop="http://www.springframework.org/schema/aop"
   xmlns:tx="http://www.springframework.org/schema/tx"
   xmlns:jaxws="http://cxf.apache.org/jaxws"
   xsi:schemaLocation="http://www.springframework.org/schema/beans
   http://www.springframework.org/schema/beans/spring-beans-3.0.xsd
   http://www.springframework.org/schema/context
   http://www.springframework.org/schema/context/spring-context-3.0.xsd
```

```
        http://www.springframework.org/schema/aop
        http://www.springframework.org/schema/aop/spring-aop-3.0.xsd
        http://www.springframework.org/schema/tx
        http://www.springframework.org/schema/tx/spring-tx-3.0.xsd
        http://cxf.apache.org/jaxws
        http://cxf.apache.org/schema/jaxws.xsd">
    <context:annotation-config />
    <!-- 使用 annotation 自动注册 bean,并保证@Required,@Autowired 的属性被注入 -->
    <context:component-scan base-package="com.qs.yzaj.service" />
    <bean
   class="org.springframework.beans.factory.config.PropertyPlaceholderConfigurer">
        <property name="locations">
            <value>/WEB-INF/jdbc.oracle.properties</value>
        </property>
     </bean>
     <bean id="oracle" destroy-method="close"
        class="org.apache.commons.dbcp.BasicDataSource">
        <property name="driverClassName" value="${jdbc.driverClassName}" />
        <property name="url" value="${jdbc.url}" />
        <property name="username" value="${jdbc.username}" />
        <property name="password" value="${jdbc.password}" />
        <property name="maxActive" value="${jdbc.maxActive}"></property>
        <property name="maxIdle" value="${jdbc.maxIdle}"></property>
        <property name="maxWait" value="${jdbc.maxWait}"></property>
     </bean>
     <bean id="dataSource" destroy-method="close"
        class="org.apache.commons.dbcp.BasicDataSource">
        <property name="driverClassName" value="${jdbc.mysql.driverClassName}" />
        <property name="url" value="${jdbc.mysql.url}" />
        <property name="username" value="${jdbc.mysql.username}" />
        <property name="password" value="${jdbc.mysql.password}"/>
        <property name="maxActive" value="${jdbc.maxActive}"></property>
        <property name="maxIdle" value="${jdbc.maxIdle}"></property>
        <property name="maxWait" value="${jdbc.maxWait}"></property>
     </bean>
     <bean id="sqlSessionFactory" class="org.mybatis.spring.SqlSessionFactoryBean">
        <property name="dataSource" ref="dataSource" />
        <property name="configLocation"
              value="classpath:mybatis/configuration.xml" />
     </bean>
     <bean id="sqlSession" class="org.mybatis.spring.SqlSessionTemplate">
        <constructor-arg index="0" ref="sqlSessionFactory" />
     </bean>
     <!-- 定义事务管理器（声明式的事务） -->
    <bean id="transactionManager"
       class="org.springframework.jdbc.datasource.DataSourceTransactionManager">
       <property name="dataSource" ref="dataSource" />
    </bean>
    <tx:annotation-driven transaction-manager="transactionManager"
       proxy-target-class="true" />
</beans>
```

8.2.4 IoC 介绍

前文已经提过，Spring 通过 IoC 容器统一管理各模块之间的依赖关系，来降低各组件之间的耦合程度，下面将具体介绍 IoC 是如何工作的。

IoC（Inverse of Control）通常称为控制反转，它是一种设计模式，主要关注组件的依赖性、配置及组件的生命周期。一般情况下，程序若需要调用某个类时，必须自己创建一个调用类的对象实例；但是采用 IoC 以后，创建对象实例的任务将由容器或框架来完成，应用程序直接使用容器或框架创建的对象即可。

控制反转实际上包括两部分内容：控制和反转。控制是指对象应该调用哪个类的控制权；反转是指控制权应该由调用对象转移到容器或框架。使用 IoC 后，对象将会被动地接受它所依赖的类，而容器会将对象依赖的类提供给它。通俗的理解是：平常我们 new 一个实例，这个实例的控制权是我们程序员，而控制反转是指 new 实例工作不由我们程序员来做而是交给 spring 容器来做。

至此，“依赖注入”的概念应运而生，所谓依赖注入是指框架或容器将被调用的类注入给调用对象，用来解除调用对象和被调用类之间的依赖关系。依赖注入有 3 种实现方式，分别为构造函数注入、设值方法注入和接口注入。

1. Set 注入

Set 注入是通过添加并使用被调用类的 setter 方法完成注入过程。下面的实例中，定义了一个 SpringAction 类，类中需要实例化一个 SpringDao 对象，并创建 SpringDao 的 set 方法。

```
package com.bless.springdemo.action;
public class SpringAction {
    private SpringDao springDao;                    //注入对象 springDao
      public void setSpringDao(SpringDao springDao) {   //注入对象的 set 方法
          this.springDao = springDao;
    }
   public void ok(){
springDao.ok();
    }
}
```

xml 文件中的配置如下：

```
<!--配置 bean,配置后该类由 spring 管理-->
    <bean name="springAction" class="com.bless.springdemo.action.SpringAction">
        <property name="springDao" ref="springDao"></property>
    </bean>
<bean name="springDao" class="com.bless.springdemo.dao.impl.SpringDaoImpl"></bean>
```

<bean>中的 name 属性是 class 属性的一个别名，class 属性指类的全名。<property>标签中的 name 就是 SpringAction 类中的 SpringDao 属性名。配置文件的意义在于：spring 将 SpringDaoImpl 对象实例化并且调用 SpringAction 的 setSpringDao 方法将 SpringDao 对象注入。

2. 构造函数注入

构造函数注入的方式是通过调用类的构造函数，并将被调用类当作参数传递给构造函数，以此实现注入。例如，下面一个实例创建了两个成员变量 SpringDao 和 User，在创建 SpringAction 对象时要将 SpringDao 和 User 两个参数值传进来：

```
public class SpringAction {
    private SpringDao springDao;                     //注入对象 springDao
    private User user;
    public SpringAction(SpringDao springDao,User user){
```

```
        this.springDao = springDao;
        this.user = user;
    }
    public void save(){
        user.setName("mm");
        springDao.save(user);
    }
}
```

xml 文件中的配置如下：

```
<!--配置 bean,配置后该类由 spring 管理-->
    <bean name="springAction" class="com.bless.springdemo.action.SpringAction">
        <constructor-arg index="0" ref="springDao"></constructor-arg>
        <constructor-arg index="1" ref="user"></constructor-arg>
    </bean>
    <bean name="springDao"
        class="com.bless.springdemo.dao.impl.SpringDaoImpl"></bean>
    <bean name="user" class="com.bless.springdemo.vo.User"></bean>
```

3. 注解实现依赖注入

通过注解实现依赖注入，首先需要在 Spring 容器的配置文件 applicationContext.Xml 中配置以下信息，该信息是一个 Spring 配置文件的模板：

```
xmlns:context="http://www.springframework.org/schema/context"
http://www.springframework.org/schema/context
http://www.springframework.org/schema/context/spring-context-2.5.xsd
```

以上的配置隐式地注册了多个对注释进行解析的处理器：AutowiredAnnotationBeanPostProcessor、CommonAnnotationBeanPostProcessor、PersistenceAnnotationBeanPostProcessor 等。

其次，在配置文件中打开<context:annotation-config>节点，告诉 Spring 容器可以用注解的方式注入依赖对象。本系统运用的即为该注解方式，其在配置文件中的代码如下：

```
<context:annotation-config />
    <!-- 使用 annotation 自动注册 bean,并保证@Required,@Autowired 的属性被注入 -->
   <context:component-scan base-package="com.qs.yzaj.service" />
```

配置文件的意义是，自动扫描“com.qs.yzaj.service”包下有没有包含注解的类，如果有，就将该类注入。

Spring 和注入相关的常见注解有 Autowired，Resource，Qualifier，Service，Controller，Repository，Component。

- Autowired：自动注入，自动从 spring 的上下文找到合适的 bean 来注入。
- Resource：指定名称注入。
- Qualifier 和 Autowired 配合使用：指定 bean 的名称。
- Service，Controller，Repository：分别标记类是 Service 层类、Controller 层类、数据存储层的类，spring 扫描注解配置时，会标记这些类要生成 bean。
- Component：是一种泛指，标记类是组件，spring 扫描注解配置时，会标记这些类要生成 bean。

下面我们通过本系统中的部分代码看一下注解注入的使用：

```
package com.qs.yzaj.service.business.enterprise;
```

```
import java.sql.SQLException;
import java.util.List;
import java.util.Map;
import org.springframework.beans.factory.annotation.Autowired;
import org.springframework.stereotype.Service;
import org.springframework.transaction.annotation.Transactional;
import com.qs.yzaj.service.business.enterprise.dao.EnterpriseInfoDao;
@Service                                              //标注@Service 注解
@Transactional(rollbackFor = { Exception.class, SQLException.class })
public class EnterpriseInfoSvc {
    @Autowired                             //通过 Autowired 修饰 EnterpriseInfoDao 字段
    EnterpriseInfoDao EnterpriseInfoDao;
    Public Object save(Map<String, Object> paramMap) throws SQLException{
        return EnterpriseInfoDao.save(paramMap);
    }
    ……
}
```

上述代码新建 EnterpriseInfoSvc 类，并给该类标注@Service 注解，在这个类中定义字段 EnterpriseInfoDao，并用 Autowired 来修饰此字段，这样之前定义的 EnterpriseInfoDao 类的实例就会自动注入到 EnterpriseInfoSvc 的实例中了。

Autowired 注解有一个可以为空的属性 required，可以用来指定字段是否是必须的，如果是必需的，则在找不到合适的实例注入时会抛出异常。

8.2.5 BeanFactory、ApplicationContext

上文中提到过 BeanFactory 是 Bean 的顶级父类，ApplicationContext 是 Context 的顶级父类。这两个接口实现了面对 IoC 的访问。

1. BeanFactory

BeanFacory 为 IoC 提供了相关的配置机制。我们知道，Spring 通过在 xml 文件中设置配置信息，可以指定对象之间的依赖关系，而在配置文件中每个对象都以 Bean 形式被配置。BeanFactory 对 Bean 组件的常见操作如下所述。

（1）创建 Bean

BeanFactory 不仅可以产生对象，也可以管理对象，其生成方式有 3 种：静态工厂创建方式、非静态工厂创建方式和构造函数方式。每种方法的配置文件写法并不相同。

静态工厂的代码示例代码：

```
<bean id="" class="工厂类" factory-method="静态工厂方法"/>
```

非静态工厂的示例代码：

```
<bean id="factory" class="工厂类"/>
    <bean id="" factory-bean="factory" factory-method="实例工厂方法"/>
```

通过构造器（有参或无参）的示例代码：

```
<bean id="" class=""/>
```

（2）初始化 JavaBean

在使用某个 Bean 组件之前，首先需要初始化一个 JavaBean 的实例对象。容器根据 xml 文件中的 Bean 组件的配置，实例化 Bean 对象，并将目标 JavaBean 注入到指定的 Bean 对象中。

（3）使用 JavaBean

一旦 JavaBean 被初始化后，就可以正常使用该实例。可以在程序中通过 getBean 方法来获得实例对象，并使用这个对象进行相关操作。

（4）销毁 JavaBean

当 Spring 的应用结束时，容器会调用相关方法来销毁已有的 JavaBean 实例。

2. ApplicationContext

ApplicationContext 继承自 BeanFactory 接口，除了包含 BeanFactory 的所有功能之外，在国际化支持、资源访问（如 URL 和文件）、事件传播等方面也进行了良好的支持。

与 BeanFactory 通常以代码的方式被创建不同，ApplicationContext 可以通过宣告的方式创建（如使用 ContextLoader 的支持类），在 Web 应用启动时自动创建 ApplicationContext。

ContextLoader 接口有 2 个实现：ContextLoaderServlet 和 ContextLoaderListener。两者在功能上完全等同，只是一个是基于 Servlet 2.3 版本中新引入的 Listener 接口实现，一个是基于 Servlet 接口实现，可以根据目标 Web 容器的实际情况进行选择。

Web.xml 中部分代码：

```
<!-- 指定 spring 配置文件位置 -->
<context-param>
    <param-name>contextConfigLocation</param-name>
    <param-value>
    <!--加载多个 spring 配置文件 -->
    WEB-INF/spring/applicationContext.xml
    </param-value>
    </context-param>
    <!-- 定义 SPRING 监听器，加载 spring -->
    <listener>
       <listener-class>org.springframework.web.context.ContextLoaderListener
       </listener-class>
    </listener>
    <listener>
         <listener-class>
         org.springframework.web.context.request.RequestContextListener
         </listener-class>
    </listener>
```

监听器首先会检查 contextConfigLocation 参数，如果不存在，则使用“/WEB-INF/applicationContext.xml”作为默认值。如果已存在，则使用设置值，用分隔符（逗号、冒号、空格）隔开。

除了 ContextLoader 接口，ApplicationContext 接口还有众多实现类，例如 ClassPathXmlApplicationContext 类，从类路径 ClassPath 中寻找指定的 XML 配置文件，找到并装载完成 ApplicationContext 的实例化工作；FileSystemXmlApplicationContext 类，从指定的文件系统路径中寻找指定的 XML 配置文件，找到并装载完成 ApplicationContext 的实例化工作。

8.3 DWR 框架介绍

DWR（Direct Web Remoting）是一个 Ajax 的开源框架，用于改善 Web 页面与 Java 类交互的

远程服务器端的交互体验，可以帮助开发人员开发包含 Ajax 技术的网站。它可以允许在前端利用 javascript 直接调用后端的 Java 方法，并返回值给 javascript，就好像直接调用一样。

本章通过工业园区安全巡检系统的代码实例，介绍 DWR 框架的配置和使用。

首先是配置 DWR 的环境。

8.3.1　配置 web.xml 文件

web.xml 文件：

```
<!-- DWR config -->
<servlet>
    <servlet-name>dwr-invoker</servlet-name>
    <servlet-class>org.directwebremoting.servlet.DwrServlet</servlet-class>
      <init-param>                                          //设置为调试模式
        <param-name>debug</param-name>
        <param-value>true</param-value>
    </init-param>
    <init-param>
        <param-name>config-1</param-name>
        <param-value>/WEB-INF/dwr/dwr-sys.xml</param-value>
    </init-param>
    <init-param>
        <param-name>config-2</param-name>
        <param-value>/WEB-INF/dwr/dwr-app.xml</param-value>
    </init-param>
    <init-param>
        <param-name>crossDomainSessionSecurity</param-name>
        <param-value>false</param-value>
    </init-param>
</servlet>
<servlet-mapping>
    <servlet-name>dwr-invoker</servlet-name>
    <url-pattern>/dwr/*</url-pattern>
    </servlet-mapping>
```

8.3.2　配置 dwr.xml 文件

DWR 中 dwr.xml 是核心配置文件，该项目中我们将 dwr-app.xml 和 dwr-sys.xml 放入 WEB-INF/dwr 文件夹下。

dwr.xml 主要的标签有：<converter>、<convert>、<create>这 3 个标签。

<converter>标签：dwr 中内置的转换器。我们也可以使用自己写的转换器，不过 dwr 提供的转换器已经足够了，所以这个标签，一般不会自己去写。

<convert>标签：将 converter 中定义的转换器映射到的具体类型。

<create>标签：dwr 中重要的标签，用来描述 Java（服务器端）与 javascript（客户端）的交互方式。

dwr-app.xml 文件：

```
<?xml version="1.0" encoding="UTF-8"?>
<dwr>
    <allow>
    <convert match="java.lang.Exception" converter="exception"/>
    <!-- 企业类型 -->
        //create 标签用来描述 java(服务器端) 与 javascript (客户端)的交互方式,其中,creator
```

```
        和 javascript 是必须属性
    <create creator="spring" javascript="EnterpriseTypeSvc">
        //通过 spring 框架访问 bean
        <param name="beanName" value="enterpriseTypeSvc" />
    </create>
    <!-- 企业信息 -->
    <create creator="spring" javascript="EnterpriseInfoSvc">
        <param name="beanName" value="enterpriseInfoSvc" />
    </create>
    <!-- 企业人员信息 -->
    <create creator="spring" javascript="EnterpriseStaffSvc">
        <param name="beanName" value="enterpriseStaffSvc" />
    </create>
    <!-- 安全级别 -->
    <create creator="spring" javascript="SecurityLevelSvc">
        <param name="beanName" value="securityLevelSvc" />
    </create>
    <!-- 拍照上传 -->
    <create creator="spring" javascript="ImagesRecordSvc">
        <param name="beanName" value="imagesRecordSvc" />
    </create>
    <!-- 巡检点信息 -->
    <create creator="spring" javascript="SupervisionSiteSvc">
        <param name="beanName" value="supervisionSiteSvc" />
    </create>
    <!-- 巡检时间安排 -->
    <create creator="spring" javascript="PatrolScheduleSvc">
        <param name="beanName" value="patrolScheduleSvc" />
    </create>
    <!-- 出勤记录 -->
    <create creator="spring" javascript="AttendanceSvc">
        <param name="beanName" value="attendanceSvc" />
    </create>
    <!-- 消息发送 -->
    <create creator="spring" javascript="MessageSvc">
        <param name="beanName" value="messageSvc" />
    </create>
    <!-- 反馈信息 -->
    <create creator="spring" javascript="FeedbackSvc">
        <param name="beanName" value="feedbackSvc" />
    </create>
    <!-- 客户端版本 -->
    <create creator="spring" javascript="ClientVersionSvc">
        <param name="beanName" value="clientVersionSvc" />
    </create>
    </allow>

    <signatures>
    <![CDATA[
            import java.util.Map;
            import java.util.List;]]>
  </signatures>
</dwr>
```

dwr-sys.xml 文件：

```
<?xml version="1.0" encoding="UTF-8"?>
<dwr>
   <allow>
      <!-- 组织管理 -->
   <create creator="spring" javascript="OrganSvc">
      <param name="beanName" value="organSvc" />
   </create>
   <!-- 人员管理 -->
   <create creator="spring" javascript="StaffSvc">
      <param name="beanName" value="staffSvc" />
   </create>
   <!-- 角色管理 -->
   <create creator="spring" javascript="RoleSvc">
      <param name="beanName" value="roleSvc" />
   </create>
   <!-- 角色操作配置 -->
   <create creator="spring" javascript="RoleOrganSvc">
      <param name="beanName" value="roleOrganSvc" />
   </create>
      <!-- 菜单管理 -->
   <create creator="spring" javascript="MenuSvc">
      <param name="beanName" value="menuSvc" />
   </create>
   <!-- 资源管理 -->
   <create creator="spring" javascript="ResourceSvc">
      <param name="beanName" value="resourceSvc" />
   </create>
   <!-- 数据字典管理 -->
   <create creator="spring" javascript="CategorySvc">
      <param name="beanName" value="categorySvc" />
   </create>
   <!-- 行政区域管理 -->
   <create creator="spring" javascript="RegionSvc">
      <param name="beanName" value="regionSvc" />
   </create>
   <!-- 岗位管理 -->
   <create creator="spring" javascript="PostSvc">
      <param name="beanName" value="postSvc" />
   </create>
   <!-- 登录管理 -->
   <create creator="spring" javascript="LoginSvc">
      <param name="beanName" value="loginSvc" />
   </create>
   <!-- 通知管理 -->
   <create creator="spring" javascript="NoteSvc">
      <param name="beanName" value="noteSvc" />
   </create>
   </allow>
</dwr>
```

8.3.3 页面配置

在 JSP 页面添加 js 文件。

```
<script type='text/javascript' src='js/util.js'></script>
<script type='text/javascript' src='js/engine.js'></script>
<script type='text/javascript' src='dwr/interface/EnterpriseInfoSvc.js'></script>
```

其中 engine.js 是必须的，如果要用到 DWR 提供的一些工具，则要引用 util.js。第三条是 DWR 自动生成的 js 文件，供前台调用，名字必须和 dwr.xml 中 create 标签的 javascript 属性值一样，且是 dwr/interface 开头的目录。接下来就可以写 java 代码了。

其他：DWR 可以设置是否采用异步方式访问 Java 代码，其代码为：

```
dwr.engine.setAsync(false);  //false 为同步，true（默认）为异步
```

8.3.4 系统代码示例

下面是一个代码中的示例，该示例是关于业务模块中企业信息部分的代码。

首先，介绍一下相关的 Java 类 EnterpriseInfoSvc，代码如下所示。

EnterpriseInfoSvc.java 所做的工作是实现将前台数据进行业务逻辑处理，如增删改查分页操作，并转入 dao 层操作。

EnterpriseInfoSvc.java：对前台数据进行业务逻辑处理

```
package com.qs.yzaj.service.business.enterprise;
import java.sql.SQLException;
import java.util.List;
import java.util.Map;
import org.springframework.beans.factory.annotation.Autowired;
import org.springframework.stereotype.Service;
import org.springframework.transaction.annotation.Transactional;
import com.qs.yzaj.service.business.enterprise.dao.EnterpriseInfoDao;

@Service
@Transactional(rollbackFor = { Exception.class, SQLException.class })
public class EnterpriseInfoSvc {
    @Autowired
    EnterpriseInfoDao EnterpriseInfoDao;
    public Object save(Map<String, Object> paramMap) throws SQLException{
     return EnterpriseInfoDao.save(paramMap);
     }
    public List<Map<String, String>> query(Map<String, Object> paramMap)
    throws SQLException{
     return EnterpriseInfoDao.query(paramMap);
     }
    public Map<String, Object> queryPagination(Map<String, Object> paramMap,
    Map<String, Object> page) throws SQLException{
      return EnterpriseInfoDao.queryPagination(paramMap,page);
     }
    public Object remove(Map<String, Object> paramMap) throws SQLException {
      return EnterpriseInfoDao.remove(paramMap);
     }
    public Object modify(Map<String, Object> paramMap) throws SQLException {
      return EnterpriseInfoDao.modify(paramMap);
     }
    public Object check(String enterpriseCode) throws SQLException{
      return EnterpriseInfoDao.check(enterpriseCode);
     }
}
```

EnterpriseInfoDao.java：访问数据库实现数据的持久化

```
package com.qs.yzaj.service.business.enterprise.dao;
import java.sql.SQLException;
import java.util.List;
import java.util.Map;
import java.util.HashMap;
import org.springframework.stereotype.Repository;
import com.qs.yzaj.model.persistence.dao.SimpleDaoSupport;
import com.qs.yzaj.model.persistence.pagiantion.Pagination;
@Repository
public class EnterpriseInfoDao extends SimpleDaoSupport{
    public Object save(Map<String, Object> paramMap) throws SQLException {
    paramMap.put("enterpriseId", getId("enterprise_id"));
    return getSqlSession().insert("enterpriseinfo.insert", paramMap);
    }
    public List<Map<String, String>> query(Map<String, Object> paramMap)
    throws SQLException{
        fuzzyQuerySupport(paramMap, "enterpriseName");
        return getSqlSession().selectList("enterpriseinfo.queryList", paramMap);
    }
    public Object queryEnterpriseCount(Map<String, Object> paramMap)
    throws SQLException {
        fuzzyQuerySupport(paramMap, "enterpriseName");
        return getSqlSession().selectOne("enterpriseinfo.queryEnterpriseCount",
        paramMap);
    }
    public Map<String, Object> queryPagination(Map<String, Object> paramMap,
    Map<String, Object> page) throws SQLException {
    Map<String, Object> retMap = new HashMap<String, Object>();
        Pagination pagination = new Pagination(page,
        queryEnterpriseCount(paramMap));
        paramMap.putAll(pagination.getParameter());
        retMap.put("data", query(paramMap));
        retMap.put("page", pagination.getPagination());
        return retMap;
    }
    public Object remove(Map<String, Object> paramMap)
     throws SQLException {
        return Integer.valueOf(getSqlSession().delete(
            "enterpriseinfo.delete", paramMap));
    }
    public Object modify(Map<String, Object> paramMap) throws SQLException {
        return Integer.valueOf(getSqlSession().update(
                    "enterpriseinfo.update", paramMap));
    }
    public Object check(String enterpriseCode) throws SQLException{
        return getSqlSession().selectList("enterpriseinfo.check", enterpriseCode);
    }
}
```

编写 JSP 页面，实现效果如下：

enterprise-info.jsp

```
<%@ page language="java" import="java.util.*" pageEncoding="utf-8"%>
<%@include file="../../../resource.jsp"%>
<!DOCTYPE HTML PUBLIC "-//W3C//DTD HTML 4.01 Transitional//EN">
<html>
```

```
  <head>
    <title>企业信息设置</title>
    <meta http-equiv="content-type" content="text/html; charset=UTF-8">
<link rel="stylesheet" type="text/css" href=
"<%= cxtPath%>/css/common/style-two-column.css">
    <script type='text/javascript'
    src='../../../dwr/interface/EnterpriseInfoSvc.js'></script>
    <script type="text/javascript" src="script/enterprise-info.js"></script>
     <!--
     <link rel="stylesheet" type="text/css" href="styles.css">
     -->
    <script>
     $(document).ready(function(){
          var w = $('body').width();
     var h = $('body').height();
     createTable(w,h-82);
     loadData();
     });
     </script>
  </head>
  <body height="100%">
    <table id="layoutTbl" width="100%" height="100%" border="0" cellspacing="0"
    cellpadding="0">
     <tr>
          <td valign="top">
          <div class="search-wrap">
          <form id="searchForm">
               <label class="form-lbl" for="enterpriseName">企业名称：</label>
               <% if(userType.equals("owner")){
                              %>
               <input type="text" class="form-txt" id="enterpriseName"
               name="enterpriseName" style="width:150px;" />  
                      <%
                            }else{
                      %>
               <input type="text" class="form-txt" id="enterpriseName"
               name="enterpriseName" value="<%= orgName%>"
                      readonly style="width:150px;" />  
                      <%
                            }
                      %>
               <label class="form-lbl"><input type="radio" name="status"
               value="1" checked/>有效</label>
               <label class="form-lbl"><input type="radio" name="status"
               value="0" />无效</label>  
               <input class="form-btn-submit" type="button"
                  value="查询" onClick="loadData()" />
                  </form>
               </div>
          </td>
     </tr>
     <tr class="bar-row">
          <td valign="top">
               <div class="bar-blue">
                    <div class="icon-table" style="float:left;">企业信息表格</div>
                    <div class="clear"></div>
```

```
            </div>
        </td>
    </tr>
    <tr>
        <td class="list-body" valign="top">
            <table id="dataGrid"></table>
        </td>
    </tr>
    <tr class="boot-row">
        <td valign="top"></td>
    </tr>
    </table>
  </body>
</html>
```

为保证程序的健壮性，对 JSP 文件中所调用的 javascript 脚本进行单独封装调用。

enterprise-info.js：

```
var formUtil = new FormUtil();
var oPage = {
    pageIndex:1,
    pageSize:20
    };
function createTable(w, h){
    var toolbar = [];
    if(gUserType =='owner'){
       toolbar = [ {
        text : '添加',
        iconCls : 'icon-add',
        handler : function() {
            doAdd();
        }
    }, '-', {
        text : '修改',
        iconCls : 'icon-edit',
        handler : function() {
            doedit();
        }
    }, '-', {
        text : '删除',
        iconCls : 'icon-remove',
        handler : function() {
            dodelete();
        }
    }];
    }
    $('#dataGrid').datagrid( $.extend(datagridOptions(), {
        width : w,
        height : h,
        fitColumns:true,
        idField : 'enterpriseId',
        columns : [ [{
        field : 'ck',
        checkbox : true
        },{
        field : 'enterpriseCode',
        title : '企业编号',
```

```
            width : 100
        },{
            field : 'enterpriseName',
            title : '企业名称',
            width : 350
        },{
            field : 'enterpriseTypeName',
            title : '企业类型',
            width : 120
        }, {
            field : 'phone',
            title : '联系电话',
            width : 100
        },{
            field : 'corporation',
            title : '法人',
            width : 80
        }, {
            field : 'charger',
            title : '负责人',
            width : 80
        }, {
            field : 'foundedDate',
            title : '成立时间',
            width : 100,
            formatter : function(value, rec) {
            return value.toString().substring(0,10);
            }
        }, {
            field : 'status',
            title : '有效标识',
            width : 100,
            formatter : function(value, rec) {
                return (value == '1') ? "有效" : "无效";
            }
        } ] ],

            onDblClickRow : function (){
            onDblClickEnterprise();
        },
            pagination:true,
            pageSize:oPage.pageSize,
            toolbar : toolbar
        }));
        var page = $('#dataGrid').datagrid('getPager');
       if (page){
          $(page).pagination({
            onSelectPage:function(pPageIndex, pPageSize){
                oPage.pageIndex = pPageIndex;
                oPage.pageSize = pPageSize;
                loadData();
            }
          });
       }
    }
    function loadData(obj){
```

```
        if(!obj)obj = formUtil.getFormValue('searchForm');
        if(obj.enterpriseName == '')delete obj.enterpriseName;
        EnterpriseInfoSvc.queryPagination(obj, oPage, function(oData){  //分页查询
                var opts = $('#dataGrid').datagrid('options');
                opts.data = {
                total:oData.page.recordCount,
              rows:oData.data
            };
            $('#dataGrid').datagrid('loadData', opts.data);
     });
     //清除选择
     $('#dataGrid').datagrid('clearSelections');
    }

    function doAdd(){
        top.openWindow({
            width:600,
            height:400,
            href:'enterprise-info-form.jsp',
            args:null,
            title:'新增企业信息'
        });
    }
    function doedit(){
        var rows = $('#dataGrid').datagrid('getChecked');
        if(rows.length<1){
              $.messager.alert('提示信息','请选择要编辑的企业信息！','info');
        return;
        }else if(rows.length>1){
              $.messager.alert('提示信息','多个企业信息被选中，请只选择一个！','info');
              return;
          }
        top.openWindow({
              width:600,
              height:400,
              href:'enterprise-info-form.jsp',
              args:rows[0],
              title:'编辑企业信息'
        });
    }
    function dodelete(){
        var ids = '';
        var rows = $('#dataGrid').datagrid('getChecked');
        if (rows.length < 1) {
              $.messager.alert('提示信息','请选择要删除的企业信息！','info');
              return;
          }
        for ( var i = 0; i < rows.length; i++) {
              ids += "," + rows[i].enterpriseId;
          }
        if(ids != '')ids = ids.substring(1);

        $.messager.confirm('确认提示', '确定要删除选中的企业信息？', function(r) {
              if (r) {
              EnterpriseInfoSvc.remove({enterpriseIds:ids},function(iRet){
                  if(iRet){
```

```
                        $.messager.alert('提示信息','删除成功！','info');
                        loadData();
                    }
                });
            }
        });
    }
    //查看企业信息详情
    function showEnterprise(){
        var rows = $('#dataGrid').datagrid('getChecked');
        if(rows.length<1){
            $.messager.alert('提示信息','请选择要查看的企业！','info');
            return;
        }else if(rows.length>1){
            $.messager.alert('提示信息','多个企业信息被选中，请只选择一条！','info');
            return;
        }
        top.openWindow({
            width:600,
            height:400,
            href:'enterprise-show-info.jsp',
            args:rows[0],
            title:'查看企业信息'
        });
    }
    function onDblClickEnterprise(){
        showEnterprise();
    }
```

在上述代码中：

（1）enterprise-info.jsp 中的代码行<%@include file="../../../resource.jsp"%>的 resource.jsp 中包含了 engine.js 的引用，再引入模块业务对应的“<script type='text/javascript' src='../../../dwr/interface/EnterpriseInfoSvc.js'></script>”保证 javascript 脚本中可以调用申明的 Java 类。

（2）enterprise-info.js 的 javascript 脚本中调用了后台 Java 类的相应方法，如 EnterpriseInfoSvc.queryPagination()方法、EnterpriseInfoSvc.remove()方法，这样就建立起了前后台数据间的交互。

8.4 系统编程实例

本小节主要针对工业园区安全巡检系统中的企业信息管理模块进行介绍。

企业信息管理模块中包含 4 个子菜单，分别为企业信息设置、企业人员管理、企业类型设置、安全级别设置 4 个子菜单，我们就其中的企业类型设置和安全级别设置两个子菜单做介绍：企业类型设置子菜单中，前端代码包括 enterprise-type.jsp、enterprise-type-form.jsp，以及单独封装的 javascript 脚本 enterprise-type.js。

enterprise-type.jsp：

//显示企业类型设置页面的主窗口，同时调用 javascript 脚本

```
<%@ page language="java" import="java.util.*" pageEncoding="utf-8"%>
```

```
<%@include file="../../../resource.jsp"%>
<!DOCTYPE HTML PUBLIC "-//W3C//DTD HTML 4.01 Transitional//EN">
<html>
  <head>
    <title>企业类型设置</title>
    <meta http-equiv="content-type" content="text/html; charset=UTF-8">
    <link rel="stylesheet" type="text/css"
    href="<%= cxtPath%>/css/common/style-two-column.css">
    <script type='text/javascript'
    src='../../../dwr/interface/EnterpriseTypeSvc.js'></script>
    <script type="text/javascript" src="script/enterprise-type.js"></script>
    <script>
     $(document).ready(function(){
          var w = $('body').width();
          var h = $('body').height();
          createTable(w,h-80);
          loadData();
     });
     </script>
  </head>
  <body>
  <table id="layoutTbl" width="100%" height="100%" border="0"
  cellspacing="0" cellpadding="0">
      <tr>
          <td valign="top">
          <div class="search-wrap">
               <form id="searchForm">
               <label class="form-lbl" for="enterpriseTypeName">企业类型: </label>
                    <input type="text" class="form-txt" id="enterpriseTypeName"
               name="enterpriseTypeName" style="width:120px;" />  
               <label class="form-lbl"><input type="radio" name="status"
               value="1" />有效</label>
               <label class="form-lbl"><input type="radio" name="status"
               value="0" />无效</label>  
                    <input class="form-btn-submit" type="button"
                    value="查询" onClick="loadData()" />
               </form>
          </div>
     </td>
     </tr>
     <tr class="bar-row">
          <td valign="top">
          <div class="bar-blue">
               <div class="icon-table" style="float:left;">企业类型表格</div>
               <div class="clear"></div>
          </div>
     </td>
     </tr>
     <tr>
     <td class="list-body" valign="top">
          <table id="dataGrid"></table>
     </td>
     </tr>
     <tr class="boot-row">
          <td valign="top"></td>
     </tr>
```

```
    </table>
  </body>
</html>
```

enterprise-type.js：

//完成主窗口表单的创建，以及数据处理函数，包括载入、新增、修改、删除。

```
 var formUtil = new FormUtil();
//创建主窗口的表单
function createTable(w, h){
    $('#dataGrid').datagrid( $.extend(datagridOptions(), {
          width : w,
          height : h,
          idField : 'enterpriseTypeId',
          columns : [ [ {
               field : 'ck',
               checkbox : true
          }, {
               field : 'enterpriseTypeName',
               title : '企业类型',
               width : 120
          }, {
               field : 'enterpriseTypeCode',
               title : '类型编号',
               width : 120

          },{
               field : 'status',
               title : '有效标识',
               width : 100,
               formatter : function(value, rec) {
                    return (value == '1') ? "有效" : "无效";
               }
          } ] ],
          pagination : false,
          toolbar : [ {
               text : '添加',
               iconCls : 'icon-add',
               handler : function() {
                    doAdd();
               }
          }, '-', {
               text : '修改',
               iconCls : 'icon-edit',
               handler : function() {
                    doedit();
               }
          }, '-', {
               text : '删除',
               iconCls : 'icon-remove',
               handler : function() {
                    dodelete();
               }
          }]
```

```
        }));
    }
    //载入数据记录
    function loadData(obj){
        if(!obj)obj = formUtil.getFormValue('searchForm');
        if(obj.enterpriseTypeName == '')delete obj.enterpriseTypeName;
        if(obj.enterpriseTypeCode == '')delete obj.enterpriseTypeCode;
        EnterpriseTypeSvc.query(obj,function(oData){                     //调用后台查询方法
            var opts = $('#dataGrid').datagrid('options');
            opts.data = {
                total:oData.length,
                rows:oData
            };
            $('#dataGrid').datagrid('loadData', opts.data);
        });
        //清除选择
        $('#dataGrid').datagrid('clearSelections');
    }
    //新增记录
    function doAdd(){
        top.openWindow({
            width:600,
            height:300,
            href:'enterprise-type-form.jsp',
            args:null,
            title:'新增企业类型'
        });
    }
    //修改记录
    function doedit(){
        var rows = $('#dataGrid').datagrid('getChecked');
        if(rows.length<1){
            $.messager.alert('提示信息','请选择要编辑的企业类型！','info');
            return;
        }else if(rows.length>1){
            $.messager.alert('提示信息','多个企业类型被选中，请只选择一个！','info');
            return;
        }
        top.openWindow({
            width:600,
            height:300,
            href:'enterprise-type-form.jsp',
            args:rows[0],
            title:'编辑企业类型'
        });
    }
    //删除记录
    function dodelete(){
        var ids = '';
        var rows = $('#dataGrid').datagrid('getChecked');
        if (rows.length < 1) {
            $.messager.alert('提示信息','请选择要删除的企业类型！','info');
            return;
```

```
    }
    for ( var i = 0; i < rows.length; i++) {
        ids += "," + rows[i].enterpriseTypeId;
    }
    if(ids != '')ids = ids.substring(1);
    $.messager.confirm('确认提示', '确定要删除选中的企业类型？', function(r) {
        if (r) {
            EnterpriseTypeSvc.remove({enterpriseTypeIds:ids},function(iRet){
                if(iRet){
                    $.messager.alert('提示信息','删除成功！','info');
                    loadData();
                }
            });
        }
    });
}
```

enterprise-type-form.jsp：

//当在主页面触发新增、修改控件时，在弹出的子窗口进行表单信息的新增或者修改操作。

```
<!DOCTYPE HTML PUBLIC "-//W3C//DTD HTML 4.01 Transitional//EN">
<%@page language="java"  contentType="text/html; charset=utf-8"%>
<%@include file="../../../resource.jsp"%>
<html>
  <head>
    <title>企业类型表单</title>
    <meta http-equiv="content-type" content="text/html; charset=UTF-8">
    <script type='text/javascript'
    src='../../../dwr/interface/EnterpriseTypeSvc.js'></script>
<style>
</style>
<script type="text/javascript">
var formUtil = new FormUtil();
var gArgs;                                          //gArgs 接收从父类窗口传过来的参数
$(document).ready(function(){
    gArgs = top.getWindowOption().args;
    init();
});
//初始化表单
function init(){
    if(gArgs){
        formUtil.setFormValue('dataForm',gArgs);
    }
}
//提交，并检查企业编号是否存在
function doSubmit(){
    var obj = formUtil.getFormValue('dataForm');
    if(!obj)return;
    if(!obj.enterpriseTypeId || obj.enterpriseTypeId == ''){//新增
          EnterpriseTypeSvc.check(obj.enterpriseTypeCode, function(iRet){
              if(iRet == 0){
                  EnterpriseTypeSvc.save(obj,function(iRet){
                      if(iRet){
                          $.messager.alert('提示信息','新增成功！','info');
```

```
                    getMainFrame().loadData();
                    top.closeWindow(false);
                    }
                });
    }else{
            $.messager.alert('提示信息','企业类型编号已经存在！','info');
        }
        });
    }else{                                                          //修改
        if(obj.enterpriseTypeCode != gArgs.enterpriseTypeCode){
            EnterpriseTypeSvc.check(obj.enterpriseTypeCode, function(iRet){
                if(iRet == 0){
                    EnterpriseTypeSvc.modify(obj,function(iRet){
                        if(iRet){
                            $.messager.alert('提示信息','修改成功！','info');
                            getMainFrame().loadData();
                            top.closeWindow(false);
                            }
                        });
                }else{
                    $.messager.alert('提示信息','企业类型编号已经存在！','info');
            }
            });
        }
    else{
        EnterpriseTypeSvc.modify(obj,function(iRet){
            if(iRet){
                $.messager.alert('提示信息','修改成功！','info',function(){
                    getMainFrame().loadData();
                    top.closeWindow(false);
                });
            }
        });
    }
    }
}
//提交
function doSubmit(){
    var obj = formUtil.getFormValue('dataForm');
    if(!obj)return;
    if(!obj.enterpriseTypeId || obj.enterpriseTypeId == ''){        //新增
        EnterpriseTypeSvc.save(obj,function(iRet){
            if(iRet){
                $.messager.alert('提示信息','新增成功！','info');
                getMainFrame().loadData();
                top.closeWindow(false);
            }
        });
    }else{                                                          //修改
        EnterpriseTypeSvc.modify(obj,function(iRet){
            if(iRet){
                $.messager.alert('提示信息','编辑成功！','info');
                getMainFrame().loadData();
                top.closeWindow(false);
```

```
                }
            });
            }
        }
    //重置
    function doReset(){
        document.getElementById("dataForm").reset();
        if(gArgs){
            formUtil.setFormValue('dataForm',gArgs);
        }
    }
    </script>
    </head>
     <body>
        <div class="dialog-body">
            <form id="dataForm" name="dataForm">
                <input type="hidden" id="enterpriseTypeId" >
                    <table class="form-table" width="100%" border="0"
                    cellspacing="0" cellpadding="2">
                     <tr>
                        <td class="form-td-blue"><label class="form-lbl"
                        for="enterpriseTypeName">企业类型：</label></td>
                        <td><input type="text" class="form-txt"  id="enterpriseTypeName"
                        name="enterpriseTypeName" maxLength=100/><span
                        class="span-red">*</span></td>
                    </tr>
                     <tr>
                        <td class="form-td-blue"><label class="form-lbl"
                        for="enterpriseTypeCode">类型编号：</label></td>
                        <td><input type="text" class="form-txt"  id="enterpriseTypeCode"
                        name="enterpriseTypeCode" maxLength=100/><span
                        class="span-red">*</span></td>
                     </tr>
                     <tr>
                        <td class="form-td-blue"><label class="form-lbl"
                        for="status">有效标识：</label></td>
                        <td>
                            <label><input type="radio" name="status"
                            value="1" checked />有效</label>
                            <label><input type="radio" name="status"
                            value="0" />无效</label>
                      </td>
                    </tr>
                </table>
                <br>
                <div align="center">
                    <input class="form-btn-submit" type="button"
                    value="提交" onClick="doSubmit()" />  
                    <input class="form-btn-cancel" type="button"
                    value="重置" onClick="doReset()">
                </div>
            </form>
        </div>
     </body>
    </html>
```

图 8.13 所示是企业类型设置菜单效果。

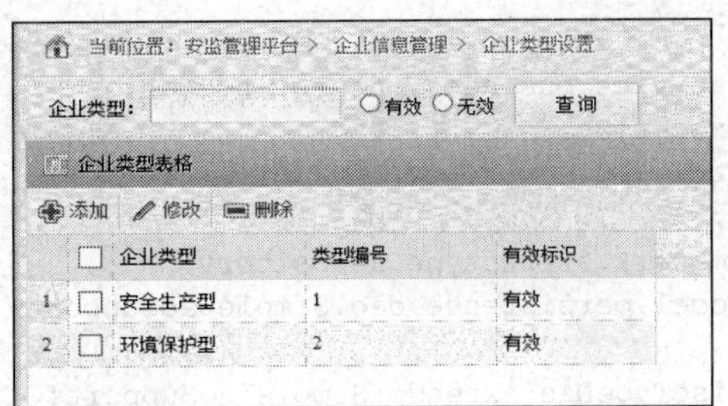

图 8.13　企业类型设置菜单效果图

后台 Java 代码包括 Service 业务层代码、dao 数据访问层代码和 sqlmap 的映射文件，分别对应于 EnterpriseTypeSvc.java、EnterpriseTypeDao.java 和 EnterpriseTypeSqlMap.xml。

其中 EnterpriseTypeSvc 类中构造了 4 个方法，save()用于新增企业类型信息，query()用于查询企业类型信息，modify()用于修改企业类型信息，remove()用于删除企业类型信息，check()用于检查企业类型编号的正确性，申明了 enterpriseTypeDao 对象，将数据处理转入 dao 层。

EnterpriseTypeSvc.java：

```
package com.qs.yzaj.service.business.enterprise;
import java.sql.SQLException;
import java.sql.SQLIntegrityConstraintViolationException;
import java.util.List;
import java.util.Map;
import org.springframework.beans.factory.annotation.Autowired;
import org.springframework.stereotype.Service;
import org.springframework.transaction.annotation.Transactional;
import com.qs.yzaj.service.business.enterprise.dao.EnterpriseTypeDao;
@Service
@Transactional(rollbackFor = { Exception.class, SQLException.class })
public class EnterpriseTypeSvc {
    @Autowired
    EnterpriseTypeDao enterpriseTypeDao;     //申明 dao 层对象
    public Object save(Map<String, Object> paramMap) throws SQLException{
        try {
            return enterpriseTypeDao.save(paramMap);
        } catch(SQLIntegrityConstraintViolationException e){
            return -1; //违反唯一性
        }
    }
    public List<Map<String, String>> query(Map<String, Object> paramMap)
    throws SQLException{
        return enterpriseTypeDao.query(paramMap);
    }
    public Object remove(Map<String, Object> paramMap) throws SQLException {
        return enterpriseTypeDao.remove(paramMap);
    }
    public Object modify(Map<String, Object> paramMap) throws SQLException {
        return enterpriseTypeDao.modify(paramMap);
    }
    public Object check(String enterpriseTypeCode) throws SQLException{
        return enterpriseTypeDao.check(enterpriseTypeCode);
    }
}
```

EnterpriseTypeDao.java：

```
package com.qs.yzaj.service.business.enterprise.dao;
import java.sql.SQLException;
import java.util.List;
import java.util.Map;
import org.springframework.stereotype.Repository;
import com.qs.yzaj.model.persistence.dao.SimpleDaoSupport;
@Repository
public class EnterpriseTypeDao extends SimpleDaoSupport{
    public Object save(Map<String, Object> paramMap) throws SQLException {
        paramMap.put("enterpriseTypeId", getId("enterprise_type_id"));
        return getSqlSession().insert("enterprisetype.insert", paramMap);
    }
    public List<Map<String, String>> query(Map<String, Object> paramMap)
    throws SQLException{
        fuzzyQuerySupport(paramMap, "enterpriseTypeName");
        return getSqlSession().selectList("enterprisetype.queryList", paramMap);
    }
    public Object remove(Map<String, Object> paramMap) throws SQLException {
        return Integer.valueOf(getSqlSession().delete(
                "enterprisetype.delete", paramMap));
    }
    public Object modify(Map<String, Object> paramMap) throws SQLException {
            return Integer.valueOf(getSqlSession().update(
                "enterprisetype.update", paramMap));
    }
    public Object check(String enterpriseTypeCode) throws SQLException{
        return getSqlSession().selectList("enterprisetype.check",
            enterpriseTypeCode);
    }
}
```

EnterpriseTypeSqlMap.xml：

//与数据库之间的映射文件，在下章中将详细介绍，一个映射文件对应于一个 dao 类

```
<?xml version="1.0" encoding="UTF-8"?>
<!DOCTYPE mapper
    PUBLIC "-//mybatis.org//DTD Mapper 3.0//EN"
    "http://mybatis.org/dtd/mybatis-3-mapper.dtd">
<mapper namespace="enterprisetype">
    <!-- 记录 -->
    <resultMap id="ResultMap" type="HashMap">
        <result property="enterpriseTypeId" column="enterprise_type_id"
        jdbcType="VARCHAR" javaType="string" />
        <result property="enterpriseTypeName" column="enterprise_type_name"
        jdbcType="VARCHAR" javaType="string" />
        <result property="enterpriseTypeCode" column="enterprise_type_code"
        jdbcType="VARCHAR" javaType="string" />
        <result property="status" column="status" jdbcType="VARCHAR"
        javaType="string" />
    </resultMap>
    <sql id="conditionSql">
        <where>
            <if test="enterpriseTypeId != null">enterprise_type_id =
            #{enterpriseTypeId}</if>
            <if test="enterpriseTypeName != null"> and enterprise_type_name like
```

```
            #{enterpriseTypeName}</if>
            <if test="enterpriseTypeCode != null"> and enterprise_type_code like
            #{enterpriseTypeCode}</if>
            <if test="status != null"> and status = #{status}</if>
        </where>
    </sql>
    <!--查询记录 -->
    <select id="queryList" parameterType="HashMap" resultMap="ResultMap">
        <include refid="pagination.paginationStart"></include>
        select enterprise_type_id, enterprise_type_name, enterprise_type_code,
            Status from enterprise_type
        <include refid="conditionSql" />
        order by enterprise_type_id desc
        <include refid="pagination.paginationLast"></include>
    </select>
    <!--新增 -->
    <insert id="insert" parameterType="HashMap">
        insert into enterprise_type(
        <trim suffixOverrides=",">
            <if test="enterpriseTypeId != null">enterprise_type_id,</if>
            <if test="enterpriseTypeName != null">enterprise_type_name,</if>
            <if test="enterpriseTypeCode != null">enterprise_type_code,</if>
            <if test="status != null">status</if>
        </trim>
        )values(
        <trim suffixOverrides=",">
            <if test="enterpriseTypeId != null">#{enterpriseTypeId},</if>
            <if test="enterpriseTypeName != null">#{enterpriseTypeName},</if>
            <if test="enterpriseTypeCode != null">#{enterpriseTypeCode},</if>
            <if test="status != null">#{status}</if>
        </trim>
        )
    </insert>
    <update id="update" parameterType="HashMap">
        update enterprise_type
        <set>
            <if test="enterpriseTypeId != null">enterprise_type_id =
                #{enterpriseTypeId},</if>
            <if test="enterpriseTypeName != null">enterprise_type_name =
                #{enterpriseTypeName},</if>
            <if test="enterpriseTypeCode != null">enterprise_type_code =
                #{enterpriseTypeCode},</if>
            <if test="status != null">status = #{status},</if>
        </set>
        where enterprise_type_id = #{enterpriseTypeId}
    </update>
    <!--删除 -->
    <delete id="delete" parameterType="HashMap">
        delete from enterprise_type where enterprise_type_id in
            (${enterpriseTypeIds})
    </delete>
    <!-- 检查企业类型编号是否已经存在 -->
    <select id="check" parameterType="HashMap" resultType="integer">
        select count(*) from enterprise_type where enterprise_type_code =
        #{enterpriseTypeCode}
    </select>
```

```
</mapper>
```

安全级别子菜单的流程与企业类型子菜单大致相同，下面是其详细代码：

security-level.jsp：

```
<%@ page language="java" import="java.util.*" pageEncoding="utf-8"%>
<%@include file="../../../resource.jsp"%>
<!DOCTYPE HTML PUBLIC "-//W3C//DTD HTML 4.01 Transitional//EN">
<html>
  <head>
    <title>安全级别设置</title>
    <meta http-equiv="content-type" content="text/html; charset=UTF-8">
    <link rel="stylesheet" type="text/css"
    href="<%= cxtPath%>/css/common/style-two-column.css">
    <script type='text/javascript'
    src='../../../dwr/interface/SecurityLevelSvc.js'></script>
    <script type="text/javascript" src="script/security-level.js"></script>
    <script type="text/javascript">
        $(document).ready(function(){
        var w = $('body').width();
        var h = $('body').height();
        createTable2(w,h-80);
        loadData();
      });
        </script>
        </head>
<body>
        <table id="layoutTbl" width="100%" height="100%" border="0"
        cellspacing="0" cellpadding="0">
         <tr>
              <td valign="top">
          <div class="search-wrap">
              <form id="searchForm">
                  <label class="form-lbl" for="securitylevelname">安全级别
                      </label>
                  <input type="text" class="form-txt" id="securitylevelname"
                      name="securitylevelname" style="width:120px;" />  
                  <label class="form-lbl"><input type="radio" name="status"
                      value="1" />有效</label>
                  <label class="form-lbl"><input type="radio" name="status"
                      value="0" />无效</label>  
                  <input class="form-btn-submit" type="button" value="查询"
                      onClick="loadData()" />
                  </form>
          </div>
          </td>
     </tr>
     <tr class="bar-row">
          <td valign="top">
          <div class="bar-blue">
          <div class="icon-table" style="float:left;">安全级别表格</div>
          <div class="clear"></div>
          </div>
          </td>
     </tr>
     <tr>
```

```
        <td class="list-body" valign="top">
            <table id="dataGrid"></table>
        </td>
    </tr>
    <tr class="boot-row">
        <td valign="top"></td>
    </tr>
    </table>
</body>
</html>
```

security-level.js：

```
var formUtil = new FormUtil();
function createTable2(w,h){
$('#dataGrid').datagrid($.extend(datagridOptions(),{
    width : w,
    height : h,
    idField : 'securitylevelid',
    columns : [[
               {
                   field : 'ck',
                   checkbox : true
               },{
                    field : 'securitylevelname',
                    title : '安全级别',
                    width : 150
               },{
                    field : 'securitylevelcode',
                    title : '级别编号',
                    width : 150
               }, {
                   field : 'status',
                   title : '有效标识',
                   width : 100,
                     formatter : function(value, rec) {
                         return (value == '1') ? "有效" : "无效";
                     }
               } ]],
               pagination : false,
                   rownumbers : true,
                   toolbar : [ {
                     text : '添加',
                     iconCls : 'icon-add',
                     handler : function() {
                         doAdd();
                     }
                 }, '-', {
                     text : '修改',
                     iconCls : 'icon-edit',
                     handler : function() {
                         domod();
                     }
                 }, '-', {
                     text : '删除',
                     iconCls : 'icon-remove',
```

```
                    handler : function() {
                        dodel();
                    }
                }]
            }));
        }
function loadData(obj)
{
    if(!obj) obj = formUtil.getFormValue('searchForm');
    if(obj.securitylevelname == '')delete obj.securitylevelname;
    SecurityLevelSvc.query(obj,function(oData){
    //   alert($d(oData));
        var opts = $('#dataGrid').datagrid('options');
        opts.data = {
            total:oData.length,
            rows:oData
           };
           $('#dataGrid').datagrid('loadData', opts.data);
   });
   //清除选择
   $('#dataGrid').datagrid('clearSelections');
   }
function doAdd(){
    top.openWindow({
        width:600,
        height:300,
        href:'security-level-form.jsp',
        args:null,
        title:'新增安全级别'
    });
}
function domod()
{
    var rows = $('#dataGrid').datagrid('getChecked');
    if(rows.length<1){
        $.messager.alert('提示信息','请选择要编辑的安全级别！','info');
        return;
    }else if(rows.length>1){
        $.messager.alert('提示信息','多个安全级别被选中，请只选择一个！','info');
        return;
    }
    top.openWindow({
        width:600,
        height:300,
        href:'security-level-form.jsp',
        args:rows[0],
        title:'编辑安全级别'
    });
}
function dodel()
{
    var ids= $('#dataGrid').datagrid('getChecked');
    if (ids.length < 1) {
        $.messager.alert('提示信息','请选择要删除的安全级别！','info');
        return;
    }
```

```
        $.messager.confirm('确认提示', '确定要删除选中的安全级别？', function(r) {
        if (r) {
            var Ids = '';
            for (var i = 0; i < ids.length; i++) {
                Ids += "," + ids[i].securitylevelid;
            }
            if(Ids != '')Ids = Ids.substring(1);
            SecurityLevelSvc.remove({securitylevelids:Ids},function(iRet){
                if(iRet){
                    $.messager.alert('提示信息','删除成功！','info');
                    loadData();
                }
            });
        }
    });
}
```

security-level-form.jsp：

```
<!DOCTYPE HTML PUBLIC "-//W3C//DTD HTML 4.01 Transitional//EN">
<%@page language="java"  contentType="text/html; charset=utf-8"%>
<%@include file="../../../resource.jsp"%>
<html>
  <head>
    <title>安全级别表单</title>
     <meta http-equiv="content-type" content="text/html; charset=UTF-8">
     <script type='text/javascript'
     src='../../../dwr/interface/SecurityLevelSvc.js'></script>
    <script type="text/javascript" src="../script/common.js"></script>
<script type="text/javascript">
var formUtil = new FormUtil();
var gArgs;//gArgs 接收从父类窗口传过来的参数
$(document).ready(function(){
    gArgs = parent.getWindowOption().args;
    init();
});
//初始化表单
function init(){
    if(gArgs){
        formUtil.setFormValue('dataForm',gArgs);
    }
}
//提交
function doSubmit(){
    var obj = formUtil.getFormValue('dataForm');
    if(!obj)return;
    if(!obj.securitylevelid || obj.securitylevelid == ''){//新增
        SecurityLevelSvc.save(obj,function(iRet){
            if(iRet > 0){
                $.messager.alert('提示信息','新增成功！','info');
                getMainFrame().loadData();
                top.closeWindow(false);
            }else if(iRet == -1){
                $.messager.alert('提示信息','级别编号重复！','info');
            }
        });
```

```
        }else{//修改
            SecurityLevelSvc.modify(obj,function(iRet){
                if(iRet > 0){
                    $.messager.alert('提示信息','编辑成功！','info');
                    getMainFrame().loadData();
                    top.closeWindow(false);
                } else if(iRet == -1){
                    $.messager.alert('提示信息','级别编号重复！','info');
                }
            });
        }
    }
    //重置
    function doReset(){
        document.getElementById("dataForm").reset();
        if(gArgs){
            formUtil.setFormValue('dataForm',gArgs);
        }
    }
    </script>
    </head>
     <body>
        <div class="dialog-body">
            <form id="dataForm" name="dataForm">
                <table class="form-table" width="100%" border="0" cellspacing="0"
                 cellpadding="2">
                  <tr>
                      <input type="hidden" id="securitylevelid" />
                    <td class="form-td-blue"><label class="form-lbl"
                     for="securitylevelname">安全级别：</label></td>
                    <td><input type="text" class="form-txt" id="securitylevelname"
                     name="securitylevelname" /><span class="span-red">*</span></td>
                  </tr>
                   <tr>
                    <td class="form-td-blue"><label class="form-lbl"
                     for="securitylevelcode">级别编号：</label></td>
                    <td><input type="text" class="form-txt"  id="securitylevelcode"
                     name="securitylevelcode" /><span class="span-red">*</span></td>
                  </tr>
                   <tr>
                    <td class="form-td-blue"><label class="form-lbl" for="status">有效
                     标识：</label></td>
                    <td>
                        <label><input type="radio" name="status" value="1" checked />
                         有效</label>
                        <label><input type="radio" name="status" value="0" />无效
                         </label>
                    </td>
                  </tr>
                </table>
                <br>
                <div align="center">
                  <input class="form-btn-submit" type="button" value="提交"
                  onClick="doSubmit()" />  
                  <input class="form-btn-cancel" type="button" value="重置"
```

```
                onClick="doReset()"/>
            </div>
        </form>
        </div>
    </body>
</html>
```

SecurityLevelSvc.java：

```
package com.qs.yzaj.service.business.enterprise;
import java.sql.SQLException;
import java.sql.SQLIntegrityConstraintViolationException;
import java.util.List;
import java.util.Map;
import org.springframework.beans.factory.annotation.Autowired;
import org.springframework.dao.DuplicateKeyException;
import org.springframework.stereotype.Service;
import org.springframework.transaction.annotation.Transactional;
import com.qs.yzaj.service.business.enterprise.dao.SecurityLevelDao;
@Service
@Transactional(rollbackFor = { Exception.class, SQLException.class })
public class SecurityLevelSvc {
    @Autowired
    SecurityLevelDao securitylevelDao;
    public Object save(Map<String, Object> paramMap) throws SQLException{
        try {
            return securitylevelDao.save(paramMap);
        } catch(DuplicateKeyException e){
            return -1; //违反唯一性
        }
    }
    public List<Map<String, String>> query(Map<String, Object> paramMap)
    throws SQLException{
        return securitylevelDao.query(paramMap);
    }
    public Object remove(Map<String, Object> paramMap) throws SQLException {
        return securitylevelDao.remove(paramMap);
    }
    public Object modify(Map<String, Object> paramMap) throws SQLException {
        try {
            return securitylevelDao.modify(paramMap);
        } catch(DuplicateKeyException e){
            return -1; //违反唯一性
        }
        }
}
```

SecurityLevelDao.java：

```
package com.qs.yzaj.service.business.enterprise.dao;
import java.sql.SQLException;
import java.util.List;
import java.util.Map;
import org.springframework.stereotype.Repository;
import com.qs.yzaj.model.persistence.dao.SimpleDaoSupport;
@Repository
public class SecurityLevelDao  extends SimpleDaoSupport
{
```

```
    public Object save(Map<String, Object> paramMap) throws SQLException {
        paramMap.put("securitylevelid", getId("security_level_id"));
        Object o = getSqlSession().insert("securitylevel.insert", paramMap);
        return o;
    }
    public List<Map<String, String>> query(Map<String, Object> paramMap)
    throws SQLException{
        return getSqlSession().selectList("securitylevel.queryList", paramMap);
    }
    public Object remove(Map<String, Object> paramMap) throws SQLException {
        return Integer.valueOf(getSqlSession().delete(
                "securitylevel.delete", paramMap));
    }
    public Object modify(Map<String, Object> paramMap) throws SQLException {
        return Integer.valueOf(getSqlSession().update(
                "securitylevel.update", paramMap));
    }
}
```

SecurityLevelSqlMap.xml：

```
<?xml version="1.0" encoding="UTF-8"?>
<!DOCTYPE mapper
    PUBLIC "-//mybatis.org//DTD Mapper 3.0//EN"
    "http://mybatis.org/dtd/mybatis-3-mapper.dtd">
<mapper namespace="securitylevel">
    <!-- 记录 -->
    <resultMap id="ResultMap" type="HashMap">
        <result property="securitylevelid" column="security_level_id"
        jdbcType="VARCHAR" javaType="string" />
        <result property="securitylevelname" column="security_level_name"
        jdbcType="VARCHAR" javaType="string" />
        <result property="securitylevelcode" column="security_level_code"
        jdbcType="VARCHAR" javaType="string" />
        <result property="status" column="status" jdbcType="VARCHAR"
        javaType="string" />
    </resultMap>
    <sql id="conditionSql">
        <where>
            <if test="securitylevelid != null">security_level_id =
            #{securitylevelid}</if>
            <if test="securitylevelname != null"> and security_level_name like
            #{securitylevelname}</if>
            <if test="securitylevelcode != null"> and security_level_code like
            #{securitylevelcode}</if>
            <if test="status != null"> and status = #{status}</if>
        </where>
    </sql>
    <!--查询记录 -->
    <select id="queryList" parameterType="HashMap" resultMap="ResultMap">
        <include refid="pagination.paginationStart"></include>
        select security_level_id, security_level_name, security_level_code, status
        from  security_level
        <include refid="conditionSql" />
        order by security_level_id DESC
        <include refid="pagination.paginationLast"></include>
    </select>
```

```
    <!--新增 -->
    <insert id="insert" parameterType="HashMap">
        insert into security_level(
        <trim suffixOverrides=",">
            <if test="securitylevelid != null">security_level_id,</if>
            <if test="securitylevelname != null">security_level_name,</if>
            <if test="securitylevelcode != null">security_level_code,</if>
            <if test="status != null">status</if>
        </trim>
        )values(
        <trim suffixOverrides=",">
            <if test="securitylevelid != null">#{securitylevelid},</if>
            <if test="securitylevelname != null">#{securitylevelname},</if>
            <if test="securitylevelcode != null">#{securitylevelcode},</if>
            <if test="status != null">#{status}</if>
        </trim>
        )
    </insert>
    <!--修改 -->
    <update id="update" parameterType="HashMap">
        update security_level
        <set>
            <if test="securitylevelid != null">security_level_id =
            #{securitylevelid},</if>
            <if test="securitylevelname != null">security_level_name =
            #{securitylevelname},</if>
            <if test="securitylevelcode != null">security_level_code =
            #{securitylevelcode},</if>
            <if test="status != null">status = #{status},</if>
        </set>
        where security_level_id = #{securitylevelid}
    </update>
    <!--删除 -->
    <delete id="delete" parameterType="HashMap">
        delete from security_level where security_level_id in (${securitylevelids})
    </delete>
</mapper>
```

图 8.14 所示是安全级别子菜单的效果图。

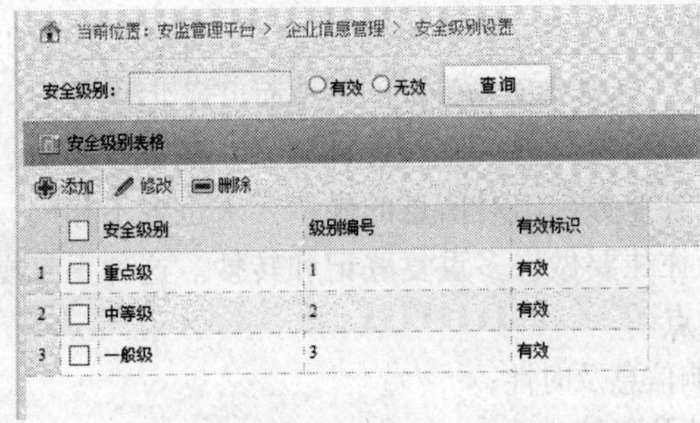

图 8.14　安全级别子菜单效果图

第9章 精细化物资与人员管理平台

9.1 平台设计

9.1.1 开发背景和需求分析

1. 开发背景

物资管理是一般工业、商业企业生产管理环节中重要的一环，物资管理的好坏，直接影响着企业的经营生产和发展。由于物资供应的渠道多、品种规格数量大，需要对物资基本信息管理、物资调配信息等进行完整的监控。加强物资管理，不但有算得出、看得到的效益，还可以大大提高管理队伍的素质，加强职工的增产节约意识、爱护财产意识和学习科技的意识，使科学管理形成共识，并可以减少物资设备各个环节上的矛盾。

企业的人员管理从一定程度上通过完善的机制和数据维护功能满足了人事部门对信息的安全及保密的特殊要求。通过精细化的物资与人员管理，可以加强企业的人员成本和产品成本的意识，能够做好物资供应、降低库存、加速资金周转、加强物资使用监督、科学化人员考核等各方面。与传统的管理方式相比，能够更合理地利用物资，完善考核机制，提高劳动生产率，促进企业健康发展。

2. 需求分析

在当今社会，互联网空间的发展，给人们的工作和生活带来了极大的便利和高效。信息化、电子化已经成为节约运营成本，提高工作效率的首选。当前大量企业的物资与员工管理尚处于手工作业阶段，不但效率低下，还常常因为管理的不慎而出现纰漏。

传统的管理方式要付出大量人力，填写各种表格、凭证、账册、卡片和文件。由于信息是随着时间不断变化的，各业务部门对信息的使用要求也各不相同，所以要按照不同的分类经常不断地汇总、统计，往往要做许多重复登记和转抄。这种手工操作的管理方式，不仅浪费人力，而且存在如下缺点。

（1）处理速度慢，影响信息及时性；

（2）易出现错误，影响信息精确性；

（3）不便于查询；

（4）缺乏综合性，不能起控制作用。

鉴于以上这些缺点，大大降低了信息的利用价值，显然越来越不适应现代物资管理工作的需要。因此，发展以电子计算机为基础的物资人员管理信息系统已是十分迫切和必要的了。该系统可以帮助企业达到员工管理办公自动化、节约管理成本、提高企业工作效率的目的。

采用精细化物资与人员管理系统，具有以下优点。

（1）利用信息技术代替人工劳动，减轻了工作量和工作的繁琐程度，降低了人员成本，提高工作质量和工作效益等。

（2）通过对库存的有效管理，可以节约资金占用，降低库存，提高经济效益。

（3）通过对材料消耗的控制，可以降低生产成本。

（4）通过管理员工的基本信息，可以达到人与事的最佳配合，根据员工的差异和表现最大程度地激励员工特质，有效管理员工。

（5）通过全面有效而准确的数据，为各层领导提供有力的决策信息。

9.1.2　系统目标与功能结构

1. 系统目标

根据系统需求分析和与客户的沟通，精细化物资与人员管理系统需要达到以下目标：

（1）系统界面设计友好、简洁大方美观；

（2）在首页中提供用户登录功能；

（3）用户登录后，系统赋予不同用户角色不同的操作权限；

（4）对人员管理方面而言，提供人员的人事信息、社保信息、劳动合同的信息的管理；

（4）对物资管理方面而言，提供物资的入库、出库、库存、采购、销售等信息的管理；

（5）统计分析模块中以图表形式展现统计的数据；

（6）系统具有易维护性和易操作性。

2. 功能结构

系统分为前、后台两部分设计，前台主要面向访问用户，实现数据信息的展示、查询与添加功能，其中信息的显示包括列表显示、图表显示和详细内容的显示。后台主要负责管理前台的数据操作。

3. 开发环境

在开发系统时，需要具备以下开发环境。

（1）服务器端

操作系统：Windows 7 操作系统。

Web 服务器：Tomcat 7.0。

Web 框架：Spring、DWR、Mybatis。

Java 开发包：JDK 1.7。

数据库：MySQL。

浏览器：IE 7.0。

分辨率：最佳效果为 1024 像素 × 768 像素。

（2）客户端

浏览器：IE 7.0。

分辨率：最佳效果为 1024 像素 × 768 像素。

9.1.3　数据库设计

本系统和上一章介绍的工业园区安全巡检管理系统一样，也是一个中小型的管理信息系统，考虑到开发成本、用户信息量及客户需求等问题，系统同样采用的是 MySQL 作为项目的数据库。

根据项目的需求，需要创建与实体对应的数据表，它们分别为数据表：用户表、部门表、驻点单位表、采购申请表、采购明细表、入库登记表、入库登记明细表、库存表、库存报警设置表、盘点记录表、出库登记表、出库等级明细表、服装发放登记表、销售明细表、数据字典表、人员基本信息表、就业登记表和参保人员登记表。

图 9.1 ~ 图 9.11 给出了其中用户表、部门表、采购申请表、采购明细表、数据字典表、库存报警设置表、入库登记表、入库登记明细表、库存表、参保人员登记表和就业登记表的表结构设计图。

用户表		
人员编号	varchar(5)	<pk>
姓名	varchar(50)	
部门编号	varchar(5)	
备注	varchar(100)	

图 9.1　用户表

部门表		
部门编号	varchar(5)	<pk>
部门姓名	varchar(50)	
上组部门编号	varchar(5)	
备注	varchar(100)	

图 9.2　部门表

采购申请表		
采购申请编号	varchar(10)	<pk>
部门编号	varchar(5)	
采购明细编号	varchar(10)	
经办人	varchar(5)	
申请日期	date	
说明问题	varchar(1000)	
部门审核人	varchar(5)	
部门审核日期	date	
部门审核意见	varchar(1000)	
财务审核人	varchar(5)	
财务审核日期	date	
财务审核意见	varchar(1000)	
领导审核	varchar(5)	
领导审核日期	date	
领导审核意见	varchar(1000)	

图 9.3　采购申请表

采购明细表		
采购明细编号	varchar(10)	<pk>
字典编号	varchar(5)	<pk>
编码	varchar(5)	<pk>
规格	varchar(5)	<pk>
品牌型号	varchar(200)	
厂家	varchar(100)	
单价	number(10,2)	
采购数量	number(5)	

图 9.4　采购明细表

数据字典表		
字典编号	varchar(5)	<pk>
字典名称	varchar(50)	
编码	varchar(5)	<pk>
编码名称	varchar(50)	

图 9.5　数据字典表

库存报警设置表		
字典编号	varchar(5)	<pk>
编码	varchar(5)	<pk>
规格	varchar(5)	<pk>
序存下限数量	number(5)	

图 9.6　库存报警设置表

入库登记表		
入库编号	varchar(10)	<pk>
入库日期	data	
入库明细编号	varchar(10)	
经办人	number(5)	
入库人	number(5)	

图 9.7　入库登记表

入库登记明细表		
入库明细编号	varchar(10)	<pk>
字典编号	varchar(5)	<pk>
编码	varchar(5)	<pk>
规格	varchar(5)	<pk>
品牌型号	varchar(200)	
厂家	varchar(100)	
入库单价	number(10, 2)	
入库数量	number(5)	

图 9.8　入库登记明细表

库存表		
库存编号	varchar(10)	<pk>
入库编号	varchar(10)	
字典编号	varchar(5)	
编码	varchar(5)	
规格	varchar(5)	
品牌型号	varchar(200)	
厂家	varchar(100)	
入库单价	number(10, 2)	
库存数量	number(5)	

图 9.9　库存表

参保人员登记表		
人员编号	varchar(20)	<pk>
社会保障卡号	varchar(100)	
姓名	varchar(20)	
性别	varchar(5)	
身份证号码	varchar(18)	
民族	varchar(5)	
户口性质	varchar(5)	
参加工作时间	date	
进本单位参保时间		
月缴费基数	number(9,2)	
备注	varchar(1000)	
工作状态	varchar(5)	

图 9.10　参保人员登记表

就业登记表		
人员序号	varchar(20)	<pk>
姓名	varchar(20)	
文化程度	varchar(5)	
劳动保障卡号	varchar(100)	
就业失业登记证号	varchar(100)	
性别	varchar(5)	
劳动合同是否签订	varchar(5)	
岗位（工种）	varchar(5)	
劳动合同起日期	date	
劳动合同止日期	date	
经办人	varchar(5)	
备注	varchar(1000)	
工作状态	varchar(5)	

图 9.11　就业登记表

9.1.4　系统预览图

图 9.12、图 9.13 给出了本章介绍的系统在实际项目投入使用后平台的预览图（部分功能模块的页面图将在后文中给出）。

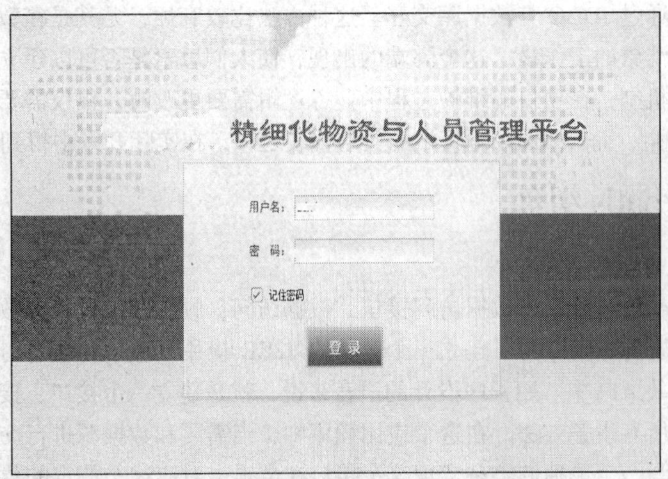

图 9.12　登录界面图

图 9.13　首页预览效果图

9.2　Mybatis 框架介绍

在本章的编程实例“精细化物资与人员管理系统”实现中，用到了实用的 Web 框架 Spring、DWR、Mybatis 和相应的技术等，由于上一章中已经介绍过 Spring 和 DWR 框架，这一章中将通过系统的代码实例介绍 Mybatis 框架技术。

9.2.1　Mybatis 概述

在 Java 项目实际开发过程，关系型数据库往往是业务模型不可或缺的一部分，传统的操纵数据库的方法是直接通过 JDBC 与数据库交互。这种方法比较麻烦，尤其是在数据量巨大时，给开发和维护带来极其浩繁的工作量，这个弊端的出现，使人们思索是否可以建立对象和关系型数据库之间的直接映射机制，在项目的维护过程中业务逻辑需要更改时，仅仅需要配置文件即可。于是 O/R 模型应运而生，而 Mybatis 是其中的典型代表之一，在实际开发中得到普遍的应用。

9.2.2　Mybatis 组件

1. DAO 组件

DAO 是 Data Access Object 数据访问接口，数据访问：顾名思义就是与数据库打交道。夹在业务逻辑与数据库资源中间。为了建立一个健壮的 J2EE 应用，应该将所有对数据源的访问操作抽象封装在一个公共 API 中。用程序设计的语言来说，就是建立一个接口，接口中定义了此应用程序中将会用到的所有事务方法。在这个应用程序中，当需要和数据源进行交互的时候，则使用这个接口，并且编写一个单独的类来实现这个接口在逻辑上对应这个特定的数据存储。

本节通过以下实例说明，Svc 调用 Dao 提供的接口，业务组件通过调用 SellInfoSvc 实现操作数据库。

SellInfoDao.java:

```
package com.yhba.service.business.sell.dao;                    //导入相关的包
import java.sql.SQLException;
import java.text.SimpleDateFormat;
import java.util.Date;
import java.util.List;
import java.util.Map;
import java.util.HashMap;
import org.apache.commons.lang.RandomStringUtils;
import org.springframework.stereotype.Repository;
import com.yhba.model.persistence.dao.SimpleDaoSupport;
import com.yhba.model.persistence.pagiantion.Pagination;

@Repository
public class SellInfoDao extends SimpleDaoSupport {
    public Object save(Map<String, Object> paramMap) throws SQLException {
        paramMap.put("sellId", getId("sell_id"));
        //按照 008(代表销售)+日期时间+3 位随机数生成申请编号
        Date now=new Date();//获取系统当前时间
       SimpleDateFormat time=new SimpleDateFormat("yyyyMMdd");      //定义格式
       String rand =RandomStringUtils.random(3, false, true);
           rand = "008"+time.format(now)+rand;
           paramMap.put("sellNo", rand);
       return getSqlSession().insert("sellinfo.insert", paramMap);
    }

public List<Map<String, String>> query(Map<String, Object> paramMap)
throws SQLException {<查询操作>
        return getSqlSession().selectList("sellinfo.queryList", paramMap);
    }

public Object querySellCount(Map<String, Object> paramMap)
throws SQLException {<查询销售数量操作>
        return getSqlSession().selectOne("sellinfo.querySellCount",   paramMap);
    }

public Map<String, Object> queryPagination(Map<String, Object> paramMap,Map<String,
Object> page) throws SQLException {<分页查询操作>
Map<String, Object> retMap = new HashMap<String, Object>();
Pagination pagination = new  Pagination(page,querySellCount(paramMap));
        paramMap.putAll(pagination.getParameter());
        retMap.put("data", query(paramMap));
        retMap.put("page", pagination.getPagination());
        return retMap;
    }

    public Object remove(Map<String, Object> paramMap) throws SQLException {
        return
    Integer.valueOf(getSqlSession().delete("sellinfo.delete",paramMap));
    }<删除操作>
```

```
    public Object modify(Map<String, Object> paramMap) throws SQLException {
        return Integer.valueOf(getSqlSession().update("sellinfo.update",paramMap));
    }<修改操作>
}
```

EmployeeInfoDao.java：

```
package com.yhba.service.business.employee.dao;
import java.sql.SQLException;
import java.util.List;
import java.util.Map;
import java.util.HashMap;
import org.springframework.stereotype.Repository;
import com.yhba.model.persistence.dao.SimpleDaoSupport;
import com.yhba.model.persistence.pagiantion.Pagination;
@Repository
public class EmployeeInfoDao extends SimpleDaoSupport {

    public Object save(Map<String, Object> paramMap) throws SQLException {
        paramMap.put("employeeId", getId("employee_id"));
        return getSqlSession().insert("employeeinfo.insert", paramMap);
    }
<查询操作>
    public List<Map<String, String>> query(Map<String, Object> paramMap)
    throws SQLException {
        fuzzyQuerySupport(paramMap, "employeeName");
        return getSqlSession().selectList("employeeinfo.queryList", paramMap);
    }
<查询数量操作>
    public Object queryEmployeeCount(Map<String, Object> paramMap)
    throws SQLException {
        return getSqlSession().selectOne("employeeinfo.queryEmployeeCount",
        paramMap);
    }
<分页查询操作>
    public Map<String, Object> queryPagination(Map<String, Object>
    paramMap,Map<String, Object> page) throws SQLException {
        Map<String, Object> retMap = new HashMap<String, Object>();
        Pagination pagination = new Pagination(page,queryEmployeeCount(paramMap));
        paramMap.putAll(pagination.getParameter());
        retMap.put("data", query(paramMap));
        retMap.put("page", pagination.getPagination());
        return retMap;
    }
<删除操作>
    public Object remove(Map<String, Object> paramMap) throws SQLException {
    return
    Integer.valueOf(getSqlSession().delete("employeeinfo.delete",paramMap));
    }
    public Object modify(Map<String, Object> paramMap) throws SQLException {
    return
    Integer.valueOf(getSqlSession().update("employeeinfo.update",paramMap));
    }
    public void updateStatus(String str) throws SQLException {
```

```
        getSqlSession().update("employeeinfo.updateStatus",str);
    }
    public Object updateRetired(Map<String, Object> paramMap) throws SQLException {
    return
  Integer.valueOf(getSqlSession().update("employeeinfo.updateRetired",paramMap));
    }
    <!更新操作>
    public List<Map<String, String>> queryStatus(Map<String, Object> paramMap)
    throws SQLException {
        return getSqlSession().selectList("employeeinfo.queryStatus", paramMap);
    }
    public List<Map<String, String>> queryUnstatus(Map<String, Object> paramMap)
    throws SQLException {
        return getSqlSession().selectList("employeeinfo.queryUnstatus", paramMap);
    }<!查询操作>
}
```

SellInfoSvc.java:

```
package com.yhba.service.business.sell;
import java.sql.SQLException;
import java.util.List;
import java.util.Map;
import org.springframework.beans.factory.annotation.Autowired;
import org.springframework.dao.DuplicateKeyException;
import org.springframework.stereotype.Service;
import org.springframework.transaction.annotation.Transactional;
import com.yhba.service.business.sell.dao.SellInfoDao;

@Service
@Transactional(rollbackFor = { Exception.class, SQLException.class })
public class SellInfoSvc {
    @Autowired
    SellInfoDao SellInfoDao;
    public Object save(Map<String, Object> paramMap) throws SQLException {
        try {
            return SellInfoDao.save(paramMap);
        } catch (DuplicateKeyException e) {
            return -1;                                  // 违反唯一性
        }
    }

    public List<Map<String, String>> query(Map<String, Object> paramMap)
    throws SQLException {<调用 SellInfoDao 查询操作>
        return SellInfoDao.query(paramMap);
    }

    public Map<String, Object> queryPagination(Map<String, Object>
    paramMap,Map<String, Object> page) throws SQLException {
      <调用 SellInfoDao 分页查询操作>
        return SellInfoDao.queryPagination(paramMap, page);
    }
      <调用 SellInfoDao 删除操作>
    public Object remove(Map<String, Object> paramMap) throws SQLException {
        return SellInfoDao.remove(paramMap);
```

```
    }
     <调用 SellInfoDao 修改操作>
    public Object modify(Map<String, Object> paramMap) throws SQLException {
        try {
            return SellInfoDao.modify(paramMap);
        } catch (DuplicateKeyException e) {
            return -1;
        }
    }
    /*public Object check(String customerCode) throws SQLException {
        return EmployeeInfoDao.check(customerCode);
    }*/
}
```

EmployeeInfoSvc.java：

```
package com.yhba.service.business.employee;
import java.io.File;
import java.io.IOException;
import java.net.URL;
import java.sql.SQLException;
import java.text.SimpleDateFormat;
import java.util.ArrayList;
import java.util.Calendar;
import java.util.Date;
import java.util.HashMap;
import java.util.List;
import java.util.Map;
import javax.servlet.ServletException;
import org.springframework.beans.factory.annotation.Autowired;
import org.springframework.dao.DuplicateKeyException;
import org.springframework.stereotype.Service;
import org.springframework.transaction.annotation.Transactional;
import com.yhba.service.business.employee.dao.EmployeeInfoDao;
import com.yhba.service.commons.CommonsUtil;
import com.yhba.service.commons.FusionChartsXMLGenerator;
@Service
@Transactional(rollbackFor = { Exception.class, SQLException.class })
public class EmployeeInfoSvc {
    @Autowired
    EmployeeInfoDao EmployeeInfoDao;
    public Object save(Map<String, Object> paramMap) throws SQLException {
        try {
            return EmployeeInfoDao.save(paramMap);
        } catch (DuplicateKeyException e) {
            return -1; // 违反唯一性
        }
    }
    public List<Map<String, String>> query(Map<String, Object> paramMap)
    throws SQLException {
        return EmployeeInfoDao.query(paramMap);
    }
    public Map<String, Object> queryPagination(Map<String, Object>
    paramMap,Map<String, Object> page) throws SQLException {
        return EmployeeInfoDao.queryPagination(paramMap, page);
    }
```

```
<!删除操作>
    public Object remove(Map<String, Object> paramMap) throws SQLException {
        return EmployeeInfoDao.remove(paramMap);
    }
<!修改操作>
    public Object modify(Map<String, Object> paramMap) throws SQLException {
            return EmployeeInfoDao.modify(paramMap);
    }
    <!修改更改为离职状态操作>
    public void updateStatus(String str) throws SQLException {
        EmployeeInfoDao.updateStatus(str);
    }
<!修改更改为退休状态操作>
    public Object updateRetired(Map<String, Object> paramMap) throws SQLException {
        return EmployeeInfoDao.updateRetired(paramMap);
    }
```

2. 配置 JDBC 属性文件 jdbc.properties

```
jdbc.mysql.driverClassName=com.mysql.jdbc.Driver
jdbc.mysql.url=jdbc\:mysql\://localhost\:3306/yhba?useUnicode\=true&characterEnco
ding\=utf-8
jdbc.mysql.username=root                                                    //用户名
jdbc.mysql.password=mysqladmin                                              //密码
jdbc.mysql.maxActive=1000
jdbc.mysql.maxIdle=20
jdbc.mysql.maxWait=3000
```

3. 配置文件 SqlmapConfig.xml

```
<?xml version="1.0" encoding="UTF-8"?>
<!DOCTYPE configuration
   PUBLIC "-//mybatis.org//DTD Config 3.0//EN"
   "http://mybatis.org/dtd/mybatis-3-config.dtd">
<configuration>
   <!-- <typeAliases alias="pagination"
type="com.yhba.model.persistence.sqlmap.pagination.PaginationSqlMap" /> -->
     <mappers>
     <mapper
resource="com/yhba/model/persistence/sqlmap/pagination/PaginationSqlMap.xml"/>
     <!-- 部门管理映射路径 -->
     <mapper
resource="com/yhba/model/persistence/sqlmap/system/SysOrganSqlMap.xml" />
     <!-- 人员管理映射路径 -->
     <mapper
resource="com/yhba/model/persistence/sqlmap/system/SysStaffSqlMap.xml" />
     <!-- 角色管理映射路径 -->
     <mapper resource="com/yhba/model/persistence/sqlmap/system/SysRoleSqlMap.xml" />
     <!-- 角色操作配置映射路径 -->
     <mapper
resource="com/yhba/model/persistence/sqlmap/system/SysRoleOrganSqlMap.xml" />
     <!-- 数据字典映射路径-->
       <mapper
resource="com/yhba/model/persistence/sqlmap/system/SysCategorySqlMap.xml" />
    <!-- 菜单管理映射路径-->
```

```
    <mapper resource="com/yhba/model/persistence/sqlmap/system/SysMenuSqlMap.xml" />
        <!-- 资源管理映射路径 -->
        <mapper
resource="com/yhba/model/persistence/sqlmap/system/SysResourceSqlMap.xml" />
        <!-- 操作管理映射路径 -->
        <mapper
resource="com/yhba/model/persistence/sqlmap/system/SysOperationSqlMap.xml" />
            <!-- 日志管理映射路径 -->
        <mapper
resource="com/yhba/model/persistence/sqlmap/system/SysLoginfoSqlMap.xml" />
        <!-- 通知管理映射路径 -->
        <mapper
resource="com/yhba/model/persistence/sqlmap/system/SysNoteSqlMap.xml" />
        <!-- 职员信息设置映射路径 -->
            <mapper
resource="com/yhba/model/persistence/sqlmap/business/EmployeeInfoSqlMap.xml" />
        <!--请假信息设置映射路径 -->
        <mapper
resource="com/yhba/model/persistence/sqlmap/business/LeaveInfoSqlMap.xml" />
        <!-- 入库登记设置映射路径 -->
        <mapper
resource="com/yhba/model/persistence/sqlmap/business/InstorageInfoSqlMap.xml" />
        <!-- 入库明细登记设置映射路径 -->
        <mapper
resource="com/yhba/model/persistence/sqlmap/business/InstorageDetailInfoSqlMap.xml" />
        <!-- 出库登记设置映射路径 -->
        <mapper
resource="com/yhba/model/persistence/sqlmap/business/OutstorageInfoSqlMap.xml" />
        <!-- 出库明细登记设置映射路径 -->
        <mapper
resource="com/yhba/model/persistence/sqlmap/business/OutstorageDetailInfoSqlMap.xml" />
        <!-- 服装发放登记设置映射路径 -->
        <mapper
resource="com/yhba/model/persistence/sqlmap/business/DressuseInfoSqlMap.xml" />
        <!-- 装备折旧比例设置映射路径 -->
        <mapper
resource="com/yhba/model/persistence/sqlmap/business/DepreInfoSqlMap.xml" />
        <!-- 库存信息设置映射路径 -->
        <mapper
resource="com/yhba/model/persistence/sqlmap/business/StockInfoSqlMap.xml" />
        <!-- 库存下限设置映射路径 -->
        <mapper
resource="com/yhba/model/persistence/sqlmap/business/StocklimitInfoSqlMap.xml" />
        <!-- 盘点映射路径 -->
        <mapper
resource="com/yhba/model/persistence/sqlmap/business/TakestockInfoSqlMap.xml" />
        <!-- 销售明细映射路径 -->
        <mapper
resource="com/yhba/model/persistence/sqlmap/business/SellInfoSqlMap.xml" />
        <!-- 公司部门设置映射路径 -->
        <mapper
resource="com/yhba/model/persistence/sqlmap/business/DepartmentInfoSqlMap.xml" />
        <!-- 采购申请映射路径 -->
        <mapper
```

```
resource="com/yhba/model/persistence/sqlmap/business/PurApplyInfoSqlMap.xml" />
        <!-- 采购明细映射路径 -->
        <mapper
resource="com/yhba/model/persistence/sqlmap/business/PurDetailInfoSqlMap.xml" />
        <!-- 商业保险设置映射路径 -->
        <mapper
resource="com/yhba/model/persistence/sqlmap/business/ComminsuInfoSqlMap.xml" />
        <!-- 医疗保险人员设置映射路径 -->
        <mapper
resource="com/yhba/model/persistence/sqlmap/business/RetiredInfoSqlMap.xml" />
        <!-- 社保关系变更申报设置映射路径 -->
        <mapper
resource="com/yhba/model/persistence/sqlmap/business/ChgeinsuInfoSqlMap.xml" />
        <!-- 工伤登记设置映射路径 -->
        <mapper
resource="com/yhba/model/persistence/sqlmap/business/InjuryInfoSqlMap.xml" />
        <!-- 参保人员设置映射路径 -->
        <mapper
resource="com/yhba/model/persistence/sqlmap/business/InsureInfoSqlMap.xml" />
        <!-- 就业登记映射路径-->
        <mapper
resource="com/yhba/model/persistence/sqlmap/business/IncontractInfoSqlMap.xml" />
        <!-- 解除合同映射路径-->
        <mapper
resource="com/yhba/model/persistence/sqlmap/business/OutcontractInfoSqlMap.xml" />
        <!-- 通知发布映射路径-->
    <mapper resource="com/yhba/model/persistence/sqlmap/business/NoticeSqlMap.xml" />
        </mappers>
    </configuration>
```

4. 属性映射文件

SQL Map XML 的根节点是 sqlMap 节点，sqlMap 根节点包括 11 个子节点，分别是 typeAlias. cacheModel.paramcterMap、resultMap、sql、statement、select、insert.Update、delete 和 procedure。每个节点里还有很多要涉及属性等的内容。这些节点的基本属性简单描述如下。

<typeAlias>元素在 SQL Map XML 文件只能定义一个，一般在当前 Mapping 映射的 javaBean 中具有唯一定义。其功能是将类别名和类全名建立一种映射关系。SqlMapConfig.xml 文件 typeAlias 节点的作用是一样的。

<cacheModel>元素是在 SQL Map XML 文件中定义的可配置缓存模式。

<parameterMap>元素是在 SQL Map XML 文件中定义的参数变量，该元素可以是 Map、Javabean 或简单数据变量。

<resultMap>元素是在 SQL Map XML 文件中定义的返回值变量，可以是 Map、Javabean 或简单数据变量。

<sql>元素是在 SQL Map XML 文件中定义的通用 SQL 语句。

<statement>元素是 SQL Map XML 文件中的通用 SQL 映射声明，可以用于任何类型的 SQL 语句。

<select>元素是在 SQL Map XML 文件中基于查询的 SQL 映射声明。

<insert>元素是在 SQL Map XML 文件中基于插入的 SQL 映射声明。

<update>元素是在 SQL Map XML 文件中基于修改的 SQL 映射声明。

<delete>元素是 SQL Map XML 文件中基于删除的 SQL 映射声明。

<procedure>元素是 SQL Map XML 文件中基于存储过程的 SQL 映射声明。

本节以销售模块 SellInfoSqlMap.xml 和 EmployeeInfoSqlMap.xml 举例，详细代码如下所示。

SellInfoSqlMap.xml：

```
<?xml version="1.0" encoding="UTF-8"?>
<!DOCTYPE mapper
    PUBLIC "-//mybatis.org//DTD Mapper 3.0//EN"
    "http://mybatis.org/dtd/mybatis-3-mapper.dtd">
<mapper namespace="sellinfo">
<!--id、result 是最简单的映射，id 为主键映射；result 是其他基本数据库表字段到实体类属性的映射。-->
<resultMap id="ResultMap" type="HashMap">
<result property="sellId" column="sell_id" jdbcType="VARCHAR"    javaType="string" />
<result property="sellNo" column="sell_no" jdbcType="VARCHAR" javaType="string
<result property="sellProduct" column="sell_product" jdbcType="VARCHAR"
javaType="string" />
<result property="sellProName" column="sellproname" jdbcType="VARCHAR"
javaType="string" /><!--销售产品名称 -->
  <result      property="sellPro1"      column="sell_pro1"      jdbcType="VARCHAR"
javaType="string" />
  <result      property="sellPro2"      column="sell_pro2"      jdbcType="VARCHAR"
javaType="string" />
<result property="sellType" column="sell_type" jdbcType="VARCHAR" javaType="string"
 /><!--销售类型 -->
<result property="sellSpec" column="sell_spec" jdbcType="VARCHAR" javaType="string"
/><!--销售种类 -->
<result property="sellFactory" column="sell_factory" jdbcType="VARCHAR" ja
vaType="string" />
<result property="sellPrice" column="sell_price" jdbcType="VARCHAR" N
javaType="string" /><!--销售价格 -->
<result property="sellCount" column="sell_count" jdbcType="VARCHAR"
javaType="string" /><!--销售数量 -->
<result property="sellStaff" column="sell_staff" jdbcType="VARCHAR"
javaType="string" /><!--销售人员 -->
<result property="sellTo" column="sell_to" jdbcType="VARCHAR" javaType="string" />
<result property="sellDate" column="sell_date" jdbcType="VARCHAR" javaType="string"
/><!--销售日期 -->
<result property="remark" column="remark" jdbcType="VARCHAR" javaType="string" />
    </resultMap>  <sql id="conditionSql">
        <where>
        <if test="sellNo!= null">sell_no = #{sellNo}</if>
        <if test="sellProduct != null"> and sell_product = #{sellProduct}</if>
        <if test="sellFactory != null"> and sell_factory = #{sellFactory}</if>
        <if test="sellStaff != null"> and sell_staff = #{sellStaff}</if>
        <if test="sellDate != null"> and sell_date = #{sellDate}</if>
        <if test="sellTo != null"> and sell_to = #{sellTo}</if>
        </where>
    </sql>
    <!--查询记录 -->
    <select id="queryList" parameterType="HashMap" resultMap="ResultMap">
        <include refid="pagination.paginationStart"></include>
        select * from vw_sell
```

```
        <include refid="conditionSql" />
        order by sell_id DESC
        <include refid="pagination.paginationLast"></include>
    </select>
    <!--查询客户记录数 -->
    <select id="querySellCount" parameterType="HashMap" resultType="integer">
        select count(*) from sell_info
         <include refid="conditionSql" />
    </select>
    <!--新增操作 -->
    <insert id="insert" parameterType="HashMap">
        insert into sell_info(sell_id,sell_no,
        <trim suffixOverrides=",">
        <if test="sellProduct != null">sell_product,</if>
        <if test="sellPro1 != null">sell_pro1,</if>
        <if test="sellPro2 != null">sell_pro2,</if>
        <if test="sellType != null">sell_type,</if>
        <if test="sellSpec != null">sell_spec,</if>
        <if test="sellFactory != null">sell_factory,</if>
        <if test="sellPrice != null">sell_price,</if>
        <if test="sellCount != null">sell_count,</if>
        <if test="sellTo != null">sell_to,</if>
        <if test="sellStaff != null">sell_staff,</if>
        <if test="sellDate != null and sellDate != '' ">sell_date,</if>
        <if test="sellDate == '' ">sell_date,</if>
        <if test="remark != null">remark</if>
        </trim>
        )values(#{sellId},#{sellNo},
        <trim suffixOverrides=",">
            <if test="sellProduct != null"> #{sellProduct},</if>
            <if test="sellPro1 != null">#{sellPro1},</if>
            <if test="sellPro2 != null">#{sellPro2},</if>
            <if test="sellType != null">#{sellType},</if>
            <if test="sellSpec != null">#{sellSpec},</if>
            <if test="sellFactory != null"> #{sellFactory},</if>
            <if test="sellPrice != null">#{sellPrice},</if>
            <if test="sellCount != null"> #{sellCount},</if>
            <if test="sellTo != null"> #{sellTo},</if>
            <if test="sellStaff != null"> #{sellStaff},</if>
            <if test="sellDate != null and sellDate != '' ">#{sellDate},</if>
            <if test="sellDate == '' ">null,</if>
           <if test="remark != null">#{remark}</if>
        </trim>
    </insert>
  <!--新增操作-->
    <update id="update" parameterType="HashMap">
        update sell_info
        <set>
        <if test="sellProduct != null">sell_product = #{sellProduct},</if>
        <if test="sellPro1 != null">sell_pro1 = #{sellPro1},</if>
        <if test="sellPro2 != null">sell_pro2 = #{sellPro2},</if>
        <if test="sellType != null">sell_type = #{sellType},</if>
        <if test="sellSpec != null">sell_spec = #{sellSpec},</if>
        <if test="sellFactory != null">sell_factory = #{sellFactory},</if>
        <if test="sellPrice != null">sell_price = #{sellPrice},</if>
```

```
            <if test="sellCount != null">sell_count = #{sellCount},</if>
            <if test="sellTo != null">sell_to = #{sellTo},</if>
            <if test="sellStaff != null">sell_staff = #{sellStaff},</if>
                <if test="sellDate != null and sellDate != '' ">sell_date =
                    #{sellDate},</if>
            <if test="sellDate == '' ">sell_date = null,</if>
            <if test="remark != null">remark = #{remark}</if>
            </set>
            where sell_id = #{sellId}
        </update>
    <!--删除操作 -->
        <delete id="delete" parameterType="HashMap">
            delete from sell_info where sell_id in (${sellIds})
        </delete>
    </mapper>
```

EmployeeInfoSqlMap.xml：

```
    <?xml version="1.0" encoding="UTF-8"?>
    <!DOCTYPE mapper
        PUBLIC "-//mybatis.org//DTD Mapper 3.0//EN"
        "http://mybatis.org/dtd/mybatis-3-mapper.dtd">
    <!--id、result 是最简单的映射，id 为主键映射；result 是其他基本数据库表字段到实体类属性的映射。-->
    <mapper namespace="employeeinfo">
        <!-- 记录 -->
        <resultMap id="ResultMap" type="HashMap">
            <result property="employeeId" column="employee_id" jdbcType="VARCHAR"
            javaType="string" />
            <result property="employeeCode" column="employee_code" jdbcType="VARCHAR"
            javaType="string" /><! 雇员编码 -->
            <result property="employeeName" column="employee_name" jdbcType="VARCHAR"
            javaType="string" />
            <result property="employeeAge" column="employee_age" jdbcType="INTEGER"
            javaType="string" /><! 雇员年龄 -->
            <result property="employeeNation" column="employee_nation"
            jdbcType="VARCHAR" javaType="string" />
            <result property="employeeType" column="employee_type" jdbcType="VARCHAR"
            javaType="string" />
            <result property="emplyeTpName" column="emplyetp_name" jdbcType="VARCHAR"
            javaType="string" /><! 雇员名字 -->
            <result property="homeAdd" column="home_add" jdbcType="VARCHAR"
            javaType="string" />
            <result property="employeeTele" column="employee_tele" jdbcType="VARCHAR"
            javaType="string" />
            <result property="employeeSex" column="employee_sex" jdbcType="VARCHAR"
            javaType="string" /><! 雇员性别 -->
            <result property="entryDate" column="entry_date" jdbcType="VARCHAR"
            javaType="string" />
            <result property="employeeStatus" column="employee_status"
            jdbcType="VARCHAR" javaType="string" />
            <result property="workstatusName" column="workstatus_name"
            jdbcType="VARCHAR" javaType="string" />
            <result property="employeeLiteracy" column="employee_literacy"
            jdbcType="VARCHAR" javaType="string" />
            <result property="literacyName" column="literacy_name" jdbcType="VARCHAR"
```

```
        javaType="string" />
        <result property="employeePolitical" column="employee_political"
        jdbcType="VARCHAR" javaType="string" />
        <result property="politicalName" column="political_name" jdbcType="VARCHAR"
        javaType="string" />
        <result property="employeeIdno" column="employee_idno" jdbcType="VARCHAR"
        javaType="string" />
        <result property="HouseholdType" column="household_type" jdbcType="VARCHAR"
        javaType="string" /><! 雇员家庭出身 -->
        <result property="housetypeName" column="housetype_name" jdbcType="VARCHAR"
        javaType="string" />
        <result property="houhldReg" column="household_register" jdbcType="VARCHAR"
        javaType="string" />
        <result property="departmentId" column="department_id" jdbcType="VARCHAR"
        javaType="string" />
        <result property="residentId" column="resident_id" jdbcType="VARCHAR"
        javaType="string" />
        <result property="departmentName" column="department_name"
        jdbcType="VARCHAR" javaType="string" />
        <result property="residentName" column="resident_name" jdbcType="VARCHAR"
        javaType="string" />
        <result property="employeeCertifiid" column="employee_certifiid"
        jdbcType="VARCHAR" javaType="string" />
        <result property="employeeHeight" column="employee_height"
        jdbcType="VARCHAR" javaType="string" />
        <result property="employeeChest" column="employee_chest" jdbcType="VARCHAR"
        javaType="string" />
        <result property="employeeWaist" column="employee_waist" jdbcType="VARCHAR"
        javaType="string" />
        <result property="employeeHead" column="employee_head" jdbcType="VARCHAR"
        javaType="string" />
        <result property="employeeShoes" column="employee_shoes" jdbcType="VARCHAR"
        javaType="string" />
        <result property="suitableSize" column="suitable_size" jdbcType="VARCHAR"
        javaType="string" />
        <result property="suitableName" column="suitable_name" jdbcType="VARCHAR"
        javaType="string" />
        <result property="looseSize" column="loose_size" jdbcType="VARCHAR"
        javaType="string" />
        <result property="looseName" column="loose_name" jdbcType="VARCHAR"
        javaType="string" />
        <result property="remark" column="remark" jdbcType="VARCHAR"
        javaType="string" />
        <result property="stopDate" column="stop_date" jdbcType="VARCHAR"
        javaType="string" />
    </resultMap>
    <sql 标签库>
    <sql id="conditionSql">
        <where>
    <if test="employeeId != null">employee_id = #{employeeId}</if>
    <if test="employeeCode != null"> and employee_code = #{employeeCode}</if>
    <if test="employeeName != null"> and employee_name like #{employeeName}</if>
    <if test="employeeAge != null"> and employee_age = #{employeeAge}</if>
    <if test="homeAdd != null"> and home_add like #{homeAdd}</if>
    <if test="employeeTele != null"> and employee_tele = #{employeeTele}</if>
```

```
<if test="employeeSex != null"> and employee_sex = #{employeeSex}</if>
<if test="entryDate != null and entryDate != '' "> and entry_date = #{entryDate}</if>
<if test="entryDate1 != null and entryDate1 != '' "> and entry_date &gt;=
#{entryDate1}</if>
<if test="entryDate2 != null and entryDate2 != '' "> and entry_date &lt;=
#{entryDate2}</if>
<if test="employeeStatus != null"> and employee_status = #{employeeStatus}</if>
<if test="departmentName != null"> and department_name = #{departmentName}</if>
<if test="residentName != null"> and resident_name = #{residentName}</if>
<if test="employeeType != null"> and employee_type = #{employeeType}</if>
<if test="employeeLiteracy != null"> and employee_literacy =
#{employeeLiteracy}</if>
<if test="employePolitical != null"> and employe_political =
#{employePolitical}</if>
<if test="employeeIdno != null"> and employee_idno = #{employeeIdno}</if>
<if test="HouseholdType != null"> and household_type = #{HouseholdType}</if>
<if test="departmentId != null"> and department_id = #{departmentId}</if>
<if test="residentId != null"> and resident_id = #{residentId}</if>
<if test="employeeCertifiid != null"> and employee_certifiid =
#{employeeCertifiid}</if>
<if test="remark != null"> and remark = #{remark}</if>
</where>
</sql>
<!--查询记录 -->
<select id="queryList" parameterType="HashMap" resultMap="ResultMap">
    <include refid="pagination.paginationStart"></include>
    select * from vw_employee
    <include refid="conditionSql" />
    order by employee_id DESC
    <include refid="pagination.paginationLast"></include>
</select>
<!--查询客户记录数 -->
<select id="queryEmployeeCount" parameterType="HashMap" resultType="integer">
    select count(*) from vw_employee
    <include refid="conditionSql" />
</select>
<!--查询入职日期及人数 -->
<select id="queryStatus" parameterType="HashMap" resultMap="statistics">
    select DATE_FORMAT(entry_date,'%Y-%m') as entry_date, COUNT(*) as
     entry_count from vw_employee where
  <if test="startTime != null and startTime != ''">DATE_FORMAT(entry_date,'%Y-%m')
  &gt;= #{startTime} and</if>
  <if test="endTime != null and endTime != ''">DATE_FORMAT(entry_date,'%Y-%m')
  &lt;= #{endTime} and</if>
  employee_status='WS_0001'
  group by DATE_FORMAT(entry_date,'%Y-%m')
</select>
<!-- 统计结果 -->
<resultMap id="statistics" type="HashMap">
    <result property="entryDate" column="entry_date" jdbcType="VARCHAR"
    javaType="string" />
    <result property="entryCount" column="entry_count" jdbcType="VARCHAR"
    javaType="string" />
</resultMap>
<!--查询离职日期及人数 -->
```

```
<select id="queryUnstatus" parameterType="HashMap" resultMap="statisticsOut">
    select DATE_FORMAT(stop_date,'%Y-%m') as stop_date, COUNT(*) as stop_count
    from vw_employee where
  <if test="startTime != null and startTime != ''">DATE_FORMAT(stop_date,'%Y-%m')
  &gt;= #{startTime} and</if>
  <if test="endTime != null and endTime != ''">DATE_FORMAT(stop_date,'%Y-%m') &lt;=
  #{endTime} and</if>
  employee_status='WS_0002'
  group by DATE_FORMAT(stop_date,'%Y-%m')
</select>
<!-- 统计结果 -->
<resultMap id="statisticsOut" type="HashMap">
    <result property="stopDate" column="stop_date" jdbcType="VARCHAR"
     javaType="string" />
    <result property="stopCount" column="stop_count" jdbcType="VARCHAR"
     javaType="string" />
</resultMap>
<!--数据插值 -->
<insert id="insert" parameterType="HashMap">
    insert into employee_info(
    <trim suffixOverrides=",">
        <if test="employeeId != null">employee_id,</if>
        <if test="employeeCode != null">employee_code,</if>
        <if test="employeeName != null">employee_name,</if>
        <if test="employeeAge != null and employeeAge != '' ">employee_age,</if>
        <if test="employeeAge == '' ">employee_age,</if>
        <if test="employeeType != null">employee_type,</if>
        <if test="homeAdd != null">home_add,</if>
        <if test="employeeTele != null">employee_tele,</if>
        <if test="employeeSex != null">employee_sex,</if>
        <if test="entryDate != null and entryDate != '' ">entry_date,</if>
        <if test="entryDate == '' ">entry_date,</if>
        <if test="remark != null">remark,</if>
        <if test="employeeStatus != null">employee_status,</if>
        <if test="employeeNation != null">employee_nation,</if>
        <if test="employeeLiteracy != null">employee_literacy,</if>
        <if test="employeePolitical != null">employee_political,</if>
        <if test="employeeIdno != null">employee_idno,</if>
        <if test="HouseholdType != null">household_type,</if>
        <if test="houhldReg != null">household_register,</if>
        <if test="departmentId != null">department_id,</if>
        <if test="residentId != null">resident_id,</if>
        <if test="employeeCertifiid != null">employee_certifiid,</if>
        <if test="employeeHeight != null">employee_height,</if>
        <if test="employeeChest != null">employee_chest,</if>
        <if test="employeeWaist != null">employee_waist,</if>
        <if test="employeeHead != null">employee_head,</if>
        <if test="employeeShoes != null">employee_shoes,</if>
        <if test="suitableSize != null">suitable_size,</if>
        <if test="looseSize != null">loose_size</if>
    </trim>
    )values(
    <trim suffixOverrides=",">
        <if test="employeeId != null">#{employeeId},</if>
```

```
            <if test="employeeCode != null">#{employeeCode},</if>
            <if test="employeeName != null"> #{employeeName},</if>
            <if test="employeeAge != null and employeeAge != ''
             ">#{employee_age},</if>
            <if test="employeeAge == '' ">null,</if>
            <if test="employeeType != null">#{employeeType},</if>
            <if test="homeAdd != null">#{homeAdd},</if>
            <if test="employeeTele != null"> #{employeeTele},</if>
            <if test="employeeSex != null">#{employeeSex},</if>
            <if test="entryDate != null and entryDate != '' "> #{entryDate},</if>
            <if test="entryDate == '' ">null,</if>
            <if test="remark != null"> #{remark},</if>
            <if test="employeeStatus != null">#{employeeStatus},</if>
            <if test="employeeNation != null">#{employeeNation},</if>
            <if test="employeeLiteracy != null"> #{employeeLiteracy},</if>
            <if test="employeePolitical != null">#{employeePolitical},</if>
            <if test="employeeIdno != null"> #{employeeIdno},</if>
            <if test="HouseholdType != null"> #{HouseholdType},</if>
            <if test="houhldReg != null"> #{houhldReg},</if>
            <if test="departmentId != null">#{departmentId},</if>
            <if test="residentId != null">#{residentId},</if>
            <if test="employeeCertifiid != null"> #{employeeCertifiid},</if>
            <if test="employeeHeight != null">#{employeeHeight},</if>
            <if test="employeeChest != null"> #{employeeChest},</if>
            <if test="employeeWaist != null"> #{employeeWaist},</if>
            <if test="employeeHead != null"> #{employeeHead},</if>
            <if test="employeeShoes != null"> #{employeeShoes},</if>
            <if test="suitableSize != null"> #{suitableSize},</if>
            <if test="looseSize != null"> #{looseSize}</if>
        </trim>
    </insert>
  <! 更新操作>
      <update id="update" parameterType="HashMap">
          update employee_info
          <set>
      <if test="employeeId != null">employee_id = #{employeeId},</if>
      <if test="employeeCode != null">employee_code = #{employeeCode},</if>
      <if test="employeeName != null">employee_name = #{employeeName},</if>
      <if test="employeeAge == ''" >employee_age = null,</if>
      <if test="employeeAge != null and employeeAge != ''" >employee_age =
       #{employeeAge},</if>
      <if test="employeeType != null " >employee_type = #{employeeType},</if>
      <if test="homeAdd != null">home_add = #{homeAdd},</if>
      <if test="employeeTele != null">employee_tele = #{employeeTele},</if>
      <if test="employeeSex != null">employee_sex = #{employeeSex},</if>
      <if test="entryDate != null and entryDate != '' ">entry_date = #{entryDate},</if>
      <if test="entryDate == '' ">entry_date = null,</if>
      <if test="remark != null">remark = #{remark},</if>
      <if test="employeeStatus != null">employee_status = #{employeeStatus},</if>
      <if test="employeeNation != null">employee_nation = #{employeeNation},</if>
      <if test="employeeLiteracy != null">employee_literacy =
      #{employeeLiteracy},</if>
       <if test="employeePolitical != null">employee_political =
       #{employeePolitical},</if>
```

```
    <if test="employeeIdno != null">employee_idno = #{employeeIdno},</if>
    <if test="HouseholdType != null">household_type = #{HouseholdType},</if>

    <if test="houhldReg != null">household_register = #{houhldReg},</if>
    <if test="departmentId != null">department_id = #{departmentId},</if>
    <if test="residentId != null">resident_id = #{residentId},</if>
    <if test="employeeCertifiid != null">employee_certifiid =
     #{employeeCertifiid},</if>
    <if test="employeeHeight != null">employee_height =
     #{employeeHeight},</if>
    <if test="employeeChest != null">employee_chest =
     #{employeeChest},</if>
    <if test="employeeWaist != null">employee_waist =
     #{employeeWaist},</if>
    <if test="employeeHead != null">employee_head = #{employeeHead},</if>
    <if test="employeeShoes != null">employee_shoes =
     #{employeeShoes},</if>
    <if test="suitableSize != null">suitable_size = #{suitableSize},</if>
    <if test="looseSize != null">loose_size = #{looseSize}</if>
        </set>
        where employee_id = #{employeeId}
    </update>
    <!--更改为离职状态-->
    <update id="updateStatus" parameterType="string">
        update employee_info set employee_status = 'WS_0002'
        where employee_code = #{value}
    </update>
    <!--更改为退休状态-->
    <update id="updateRetired" parameterType="string">
        update employee_info set employee_status = 'WS_0003'
        where employee_code = #{value}
    </update>
    <!--删除操作 -->
    <delete id="delete" parameterType="HashMap">
        delete from employee_info where employee_id in (${employeeIds})
    </delete>
    <!-- 检查企业编号是否已经存在 -->
   <select id="check" parameterType="HashMap" resultType="integer">
       select count(*) from employee_info where employee_code = #{employeeCode}
  </select>
</mapper>
```

9.3　系统编程实例

本小节主要对管理平台中涉及的劳动合同管理模块进行介绍：劳动合同管理模块中包含了两个子菜单，分别为合同签订登记和劳动解除登记。

图 9.14 中，左边为业务模块的菜单栏，可以看出，每个模块都对应于各自不同的子菜单。合同签订登记子菜单的业务逻辑包含了企业所签订的合同内容的显示，同时提供查询、新增、修改、删除等基本功能。

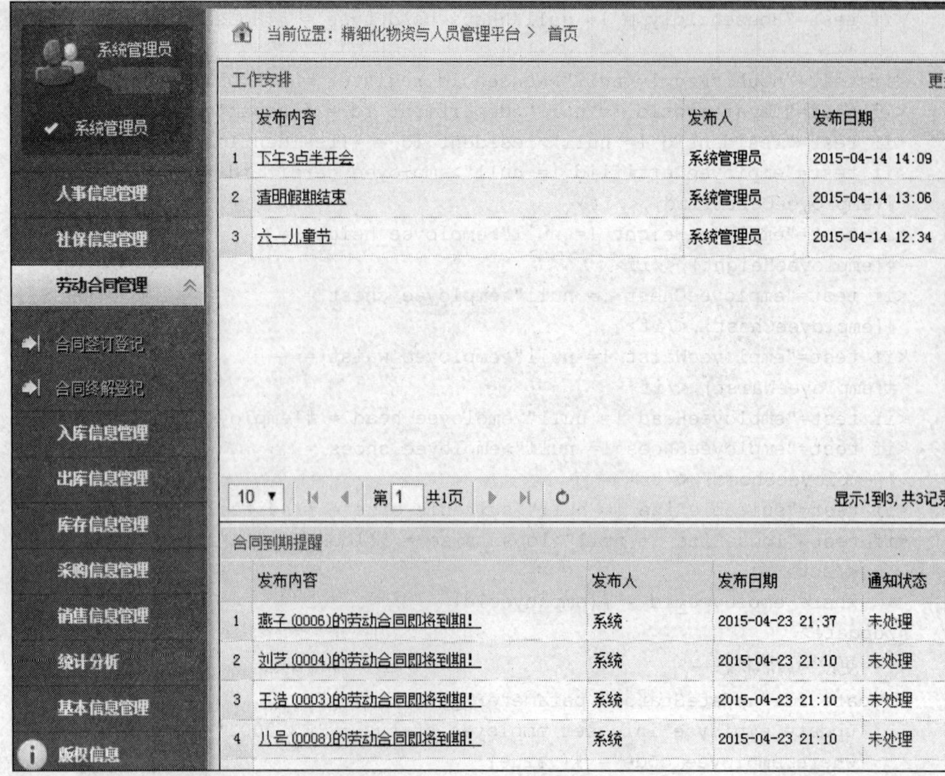

图 9.14　劳动合同模块及其子菜单

前台页面显示的 JSP 代码包括 incontract-info.jsp、incontract-form.jsp、incontract-show.jsp 和封装后的 javascript 脚本 incontract-info.js。

incontract-info.jsp：

```
<%@ page language="java" import="java.util.*" pageEncoding="utf-8"%>
<%@include file="../../../resource.jsp"%>
<%
    String roleCode = session.getAttribute("roleCode").toString();
%>
<!DOCTYPE HTML PUBLIC "-//W3C//DTD HTML 4.01 Transitional//EN">
<html>
  <head>
    <title>就业信息设置</title>
    <meta http-equiv="content-type" content="text/html; charset=UTF-8">
    <script type='text/javascript' src='../../../dwr/interface/IncontractInfoSvc.js'>
</script>
    <script type="text/javascript" src="scripts/incontract-info.js"></script>
<script>
var roleCode ='<%= roleCode %>';
    $(document).ready(function(){
        var w = $('body').width();
        var h = $('body').height();
        createTable(w,h-40);
        if(roleCode=="leader"){
            $('div.dataGrid-toolbar a').hide();//隐藏 toolbar
            $('div.datagrid-toolbar div').hide();
```

```
        }
        loadData();
        });
        </script>
    </head>
    <body height="100%">
        <table id="layoutTbl" width="100%" height="100%" border="0" cellspacing="0"
        cellpadding="0">
        <tr>
            <td valign="top">
                <div class="search-wrap">
                    <form id="searchForm">
                    <label class="form-lbl" for="employeeCode">人员胸号: </label>
                        <input type="text" class="form-txt" id="employeeCode"
name="employeeCode" style="width:120px;" />  
                    <label class="form-lbl" for="securityNo">劳动保障卡号: </label>
                        <input type="text" class="form-txt" id="securityNo"
name="securityNo" style="width:120px;" />  
                        <input class="form-btn-submit" type="button" value="查询"
onClick="loadData()" />
                    </form>
                    <table id="dataGrid"></table>
                </div>
            </td>
        </tr>
    </table>
    </body>
</html>
```

incontract-form.jsp:

```
<!DOCTYPE HTML PUBLIC "-//W3C//DTD HTML 4.01 Transitional//EN">
<%@page language="java"  contentType="text/html; charset=utf-8"%>
<%@page import="org.springframework.context.ApplicationContext"%>
<%@page
import="org.springframework.web.context.support.WebApplicationContextUtils"%>
<%@page import="com.yhba.service.system.category.CategorySvc" %>
<%@page import="java.util.*;"%>
<%@include file="../../../resource.jsp"%>
<%ApplicationContext ctx=
WebApplicationContextUtils.getWebApplicationContext(request.getSession().getServle
tContext());
   CategorySvc categorySvc = (CategorySvc) ctx.getBean("categorySvc");
  List<Map<String,String>> jobList = categorySvc.queryCategoryList("JOBTYPE");%>
<html>
  <head>
    <title>员工信息表单</title>
      <meta http-equiv="content-type" content="text/html; charset=UTF-8">
      <script type='text/javascript'
      src='../../../dwr/interface/IncontractInfoSvc.js'></script>
      <script type="text/javascript"
      src="../../../js/calendar/WdatePicker.js"></script>
      <script type='text/javascript'
      src='../../../dwr/interface/CategorySvc.js'></script>
<script type="text/javascript">
var formUtil = new FormUtil();
```

```
var gArgs;//gArgs 接收从父类窗口传过来的参数
$(document).ready(function(){
    gArgs = top.getWindowOption().args;
    init();
});
//初始化表单
function init(){
    if(gArgs){
        formUtil.setFormValue('dataForm',gArgs);
    }
}
//提交
function doSubmit(){
    var obj = formUtil.getFormValue('dataForm');
    if(!obj.incontractId || obj.incontractId == ''){                    //新增
        IncontractInfoSvc.save(obj,function(iRet){
            if(iRet > 0){
                $.messager.alert('提示信息','新增成功！','info');
                getMainFrame().loadData();
                top.closeWindow(false);
            }else if(iRet == -1){
                $.messager.alert('提示信息','已登记过此人员或此劳动保障卡号！','info');
            }
        });
    }else{                                                              //修改
        IncontractInfoSvc.modify(obj,function(iRet){
            if(iRet > 0){
                $.messager.alert('提示信息','编辑成功！','info');
                getMainFrame().loadData();
                top.closeWindow(false);
            } else if(iRet == -1){
                $.messager.alert('提示信息','已登记过此人员或此劳动保障卡号！','info');
            }
        });
    }
}
//重置
function doReset(){
    document.getElementById("dataForm").reset();
    window.location.reload();
}
</script>
</head>
 <body>
    <div class="dialog-body">
        <form id="dataForm" name="dataForm">
          <input type="hidden" id="incontractId"/>
            <table class="form-table" width="100%" border="0" cellspacing="0"
             cellpadding="5"  >
                <tr>
                    <td class="form-td-blue"  height=40><label class="form-lbl"
                    for="employeeCode">人员胸号：</label></td>
                          <td ><input type="text" class="form-txt"  id="employeeCode"
name="employeeCode"  height=100%  /><span class="span-red">*</span></td>
```

```
            </tr>
            <tr>
                <td  class="form-td-blue"   height=40><label  class="form-lbl"
for="employeeName">人员姓名: </label></td>
                <td  ><input  type="text"  class="form-txt"    id="employeeName"
name="employeeName"  height=100%  /><span class="span-red">*</span></td>
            </tr>
            <tr>
                <td    class="form-td-blue"height=40><label     class="form-lbl"
for="securityNo">劳动保障卡号: </label></td>
                    <td  ><input  type="text"  class="form-txt"    id="securityNo"
name="securityNo"   height=100% /><span class="span-red">*</span></td>
            </tr>
            <tr>
                <td    class="form-td-blue"height=40><label     class="form-lbl"
for="jobCategory">工种: </label></td>
                <td >
                 <select      class="easyui-combobox"              id="jobCategory"
name="jobCategory"  style="width:180px; border:1px solid #DCDCDC;">
                        <option value=" ">--选择工种--</option>
                        <% for(int i=0; i<jobList.size(); i++){
                            Map jobMap = (Map)jobList.get(i);
                        %>
                        <option value="<%= jobMap.get("value")%>"><%=
                         jobMap.get("text")%></option>
                        <%
                            }
                        %>
                      </select>
                </td>
            </tr>
           <tr>
                    <td   class="form-td-blue"height=40><label    class="form-lbl"
for="registerNo">就业失业登记证号: </label></td>
                    <td><input   type="text"   class="form-txt"    id="registerNo"
name="registerNo"  /><span class="span-red">*</span></td>
            </tr>
            <tr>
                <td class="form-td-blue"height=40><label class="form-lbl"
              for="signUp">是否签订劳动合同: </label></td>
                    <td>
                        <label><input type="radio" name="signUp" value="1"
                         checked />是</label>
                       <label><input type="radio" name="signUp" value="0" />否
                        </label>
                     </td>
            </tr>
            <tr>
                <td class="form-td-blue"><label class="form-lbl" for="signTime">
劳动合同签订日期: </label></td>
                    <td><input type="text" class="Wdate" style="height:30px;"
id="signTime"name="signTime"
onfocus="WdatePicker({skin:'whyGreen',dateFmt:'yyyy-MM-dd',maxDate:'%y-%M-%d',highLine
WeekDay:true})" /></td>
        </tr>
```

```
        <tr>
            <td class="form-td-blue"><label class="form-lbl" for="renewTime">劳
动合同续签日期: </label></td>
                <td><input type="text" class="Wdate" style="height:30px;"
id="renewTime" name="renewTime"
onfocus="WdatePicker({skin:'whyGreen',dateFmt:'yyyy-MM-dd',maxDate:'%y-%M-%d',highLine
WeekDay:true})" /></td>
        </tr>
        <tr>
            <td class="form-td-blue"><label class="form-lbl" for="startTime">劳
动合同起始日期: </label></td>
                <td><input  type="text"  class="Wdate"  style="height:30px;"
id="startTime" name="startTime"onFocus="var
stopTime=$dp.$('stopTime');WdatePicker({skin:'whyGreen',minDate:'#F{$dp.$D(\'signTime\
')}',maxDate:'#F{$dp.$D(\'stopTime\')}',dateFmt:'yyyy-MM-dd',highLineWeekDay:true})"/>
<span class="span-red">*</span></td>
        </tr>
         <tr>
             <td class="form-td-blue"><label class="form-lbl"
for="stopTime">劳动合同终止日期: </label></td>
             <td><input type="text" class="Wdate" style="height:30px;"
id="stopTime" name="stopTime"
onfocus="WdatePicker({skin:'whyGreen',dateFmt:'yyyy-MM-dd',minDate:'#F{$dp.$D(\'startT
ime\')}',highLineWeekDay:true})" /><span class="span-red">*</span></td>
           </tr>
           <tr>
             <td class="form-td-blue"height=40><label class="form-lbl"
for="operator">经办人: </label></td>
             <td><input type="text" class="form-txt"  id="operator"
name="operator"  /></td>
           </tr>
         <tr>
             <td class="form-td-blue"><label class="form-lbl" for="remark">备
注: </label></td>
           <td colspan="3"><textarea class="form-txt" maxLength=2000
id="remark" name="remark" rows="3" style="width:80%;height:50px;"></textarea></td>
        </tr>
      </table>
      <br>
      <div align="center">
         <input class="form-btn-submit" type="button" value="提交"
onClick="doSubmit()" />  
         <input class="form-btn-cancel" type="button" value="重置"
onClick="doReset()">
        </div>
      </form>
   </div>
 </body>
</html>
```

incontract-show.jsp:

```
<%@ page language="java"  contentType="text/html; charset=utf-8"
import="java.util.*"%>
<%@page import="org.springframework.context.ApplicationContext"%>
<%@page
import="org.springframework.web.context.support.WebApplicationContextUtils"%>
```

```
<%@include file="../../../resource.jsp"%>
<!DOCTYPE HTML PUBLIC "-//W3C//DTD HTML 4.01 Transitional//EN">
<html>
  <head>
    <title>客户信息</title>
    <meta http-equiv="content-type" content="text/html; charset=UTF-8">
<script type="text/javascript" src="../../../js/calendar/WdatePicker.js"></script>
    <script type='text/javascript'
src='../../../dwr/interface/EmployInfoSvc.js'></script>
<script type="text/javascript">
var formUtil = new FormUtil();
var gArgs;//gArgs 接收从父类窗口传过来的参数
$(document).ready(function(){
    gArgs = top.getWindowOption().args;
    init();
});
function init(){
    if(gArgs){
        formUtil.setFormValue('dataForm',gArgs);
    }
}
</script>
</head>
<body>
    <div class="dialog-body">
        <form id="dataForm" name="dataForm">
          <input type="hidden" id="incontractId"/>
            <table class="form-table" width="100%" border="0" cellspacing="0"
cellpadding="5" >
            <tr>
                <td class="form-td-blue"  height=40><label class="form-lbl"
for="employeeCode">人员胸号: </label></td>
                <td ><input disabled type="text" class="form-txt"
id="employeeCode" name="employeeCode"  height=100% /></td>
            </tr>
             <tr>
              <td class="form-td-blue"  height=40><label class="form-lbl"
for="employeeName">人员姓名: </label></td>
                  <td ><input disabled type="text" class="form-txt"
id="employeeName" name="employeeName"  height=100%  /></td>
                </tr>
            <tr>
                <td class="form-td-blue"height=40><label class="form-lbl"
for="securityNo">劳动保障卡号: </label></td>
                    <td ><input disabled type="text"
class="form-txt"  id="securityNo"name="securityNo"  height=100% /></td>
            </tr>
            <tr>
                <td class="form-td-blue"height=40><label class="form-lbl"
for="jobcateName">工种: </label></td>

                  <td ><input disabled type="text" class="form-txt"
id="jobcateName" name="jobcateName" /></td>
            </tr>
            <tr>
                <td class="form-td-blue"height=40><label class="form-lbl"
```

```
for="registerNo">就业失业登记证号：</label></td>
                <td><input disabled type="text" class="form-txt"
id="registerNo" name="registerNo"  /></td>
            </tr>
            <tr>
                <td class="form-td-blue"height=40><label class="form-lbl"
for="signUp">是否签订劳动合同：</label></td>
                <td>
                    <label><input disabled type="radio" name="signUp"
value="1" checked />是</label>
                    <label><input disabled type="radio" name="signUp"
value="0" />否</label>
                </td>
            </tr>
            <tr>
                <td class="form-td-blue"><label class="form-lbl" for="signTime">劳动
合同签订日期：</label></td>
                <td><input disabled type="text" style="height:30px;"
class="Wdate" id="signTime" name="signTime"
onfocus="WdatePicker({skin:'whyGreen',dateFmt:'yyyy-MM-dd',maxDate:'%y-%M-%d',highLine
WeekDay:true})" /></td>
            </tr>
            <tr>
                <td class="form-td-blue"><label class="form-lbl" for="renewTime">劳
动合同续签日期：</label></td>
                <td><input disabled type="text" style="height:30px;"
class="Wdate" id="renewTime" name="renewTime"
onfocus="WdatePicker({skin:'whyGreen',dateFmt:'yyyy-MM-dd',maxDate:'%y-%M-%d',highLine
WeekDay:true})" /></td>
            </tr>
            <tr>
                <td class="form-td-blue"><label class="form-lbl"
for="startTime">劳动合同起始日期：</label></td>
                <td><input disabled type="text" class="Wdate"
style="height:30px;" id="startTime" name="startTime"
onfocus="WdatePicker({skin:'whyGreen',dateFmt:'yyyy-MM-dd',maxDate:'%y-%M-%d',highLine
WeekDay:true})" /></td>
                </tr>
                <tr>
                    <td class="form-td-blue"><label class="form-lbl" for="stopTime">
劳动合同终止日期：</label></td>
                <td><input disabled type="text" class="Wdate"
style="height:30px;" id="stopTime" name="stopTime"
onfocus="WdatePicker({skin:'whyGreen',dateFmt:'yyyy-MM-dd',maxDate:'%y-%M-%d',highLine
WeekDay:true})" /></td>
            </tr>
            <tr>
                <td class="form-td-blue"height=40><label class="form-lbl"
for="operator">经办人：</label></td>
                    <td><input disabled type="text" class="form-txt"
id="operator" name="operator"  /></td>
            </tr>
            <tr>
                <td class="form-td-blue"><label class="form-lbl" for="remark">备
注：</label></td>
                <td colspan="3"><textarea disabled class="form-txt"
```

```
maxLength=2000 id="remark" name="remark" rows="3"
style="width:95%;height:50px;"></textarea></td>
                </tr>
                 </table>
                 <br>
                 <div align="center">
                    <input class="form-btn-submit" type="button" value="关闭"
onClick="top.closeWindow()" />
                 </div>
              </form>
         </div>
    </body>
    </html>
```

incontract-info.js:

```
var formUtil = new FormUtil();
var oPage = {
        pageIndex:1,
        pageSize:20
    };
function createTable(w, h){
    $('#dataGrid').datagrid( $.extend(datagridOptions(), {
        width : w,
        height : h,
        fitColumns:true,
        idField : 'incontractId',
        columns : [ [{
            field : 'ck',
            checkbox : true
        },{
            field : 'employeeCode',
            title : '人员胸号',
            width :100
        },{
            field : 'employeeName',
            title : '人员姓名',
            width :100
        },{
            field : 'securityNo',
            title : '劳动保障卡号',
            width : 100

        },{
            field : 'jobcateName',
            title : '工种',
            width : 80
        },{
            field : 'registerNo',
            title : '就业失业登记证号',
            width : 100,
        }, {
            field : 'signUp',
            title : '是否签订劳动合同',
            width : 80,
```

```
            formatter : function(value, rec) {
                return (value == '1') ? "是" : "否";
            }
        },{
            field : 'startTime',
            title : '劳动合同起始日期',
            width : 80
        },{
            field : 'stopTime',
            title : '劳动合同终止日期',
            width : 80
        },{
            field : 'operator',
            title : '经办人',
            width : 80
        } ] ],
        onDblClickRow : function (){
            onDblClick();
        },
        pagination:true,
        pageSize:oPage.pageSize,
        toolbar : [ {
            text : '添加',
            iconCls : 'icon-add',
            handler : function() {
                doAdd();
            }
        }, '-', {
            text : '修改',
            iconCls : 'icon-edit',
            handler : function() {
                doedit();
            }
        }, '-', {
            text : '删除',
            iconCls : 'icon-remove',
            handler : function() {
                dodelete();
            }
        }]
    }));
    var page = $('#dataGrid').datagrid('getPager');
    if (page){
        $(page).pagination({
            onSelectPage:function(pPageIndex, pPageSize){
                oPage.pageIndex = pPageIndex;
                oPage.pageSize = pPageSize;
                loadData();
            }
        });
    }
}
function loadData(obj){
    if(!obj)obj = formUtil.getFormValue('searchForm');
```

```
    if(obj.employeeCode == '')delete obj.employeeCode;
    if(obj.securityNo == '')delete obj.securityNo;
    if(obj.employeeCode||obj.securityNo){
    //在其他 page 下按搜索条件查询，需要初始化 pageIndex
        oPage = {
                pageIndex:1,
                pageSize:20
            };
    }
    IncontractInfoSvc.queryPagination(obj, oPage, function(oData){  //分页查询
        var opts = $('#dataGrid').datagrid('options');
        opts.data = {
            total:oData.page.recordCount,
        rows:oData.data
        };
        $('#dataGrid').datagrid('loadData', opts.data);
  });
  //清除选择
  $('#dataGrid').datagrid('clearSelections');
}
function doAdd(){
    top.openWindow({
        width:400,
        height:600,
        href:'incontract-form.jsp',
        args:null,
        title:'新增就业登记信息'
    });
}
function doedit(){
    var rows = $('#dataGrid').datagrid('getChecked');
    if(rows.length<1){
        $.messager.alert('提示信息','请选择要编辑的信息！','info');
        return;
    }else if(rows.length>1){
        $.messager.alert('提示信息','多条信息被选中，请只选择一个！','info');
        return;
    }
    top.openWindow({
        width:400,
        height:600,
        href:'incontract-form.jsp',
        args:rows[0],
        title:'编辑就业登记信息'
    });
}
function dodelete(){
    var ids = '';
    var rows = $('#dataGrid').datagrid('getChecked');
    if (rows.length < 1) {
        $.messager.alert('提示信息','请选择要删除的员工信息！','info');
        return;
    }
    for ( var i = 0; i < rows.length; i++) {
```

```
            ids += "," + rows[i].employId;
        }
        if(ids != '')ids = ids.substring(1);
        $.messager.confirm('确认提示', '确定要删除选中的员工信息？', function(r) {
            if (r) {
                IncontractInfoSvc.remove({employIds:ids},function(iRet){
                    if(iRet){
                        $.messager.alert('提示信息','删除成功！','info');
                        loadData();
                    }
                });
            }
        });
    }
    function showContract(){
        var rows = $('#dataGrid').datagrid('getChecked');
        if(rows.length<1){
            $.messager.alert('提示信息','请选择要查看的员工！','info');
            return;
        }else if(rows.length>1){
            $.messager.alert('提示信息','多条员工信息被选中，请只选择一条！','info');
            return;
        }
        top.openWindow({
            width:400,
            height:600,
            href:'incontract-show.jsp',
            args:rows[0],
            title:'查看就业登记信息'
        });
    }
    function onDblClick(){
        showContract();
    }
```

后台 Java 代码包括 service 业务层代码、dao 数据访问层代码和 sqlmap 的映射文件，分别对应于 IncontractInfoSvc.java、IncontractInfoDao.java 和 IncontractInfoSqlMap.xml。

IncontractInfoSvc.java：

（主要包括 6 种方法，save()用于保存合同签订记录、query()用于查询合同签订记录、queryPagination()用于分页查询、modify()用于修改合同签订记录、remove()用于删除合同签订记录、queryEnd()查询合同是否到期。）

```
package com.yhba.service.business.contract;
import java.sql.SQLException;
import java.util.ArrayList;
import java.util.HashMap;
import java.util.List;
import java.util.Map;
import org.springframework.beans.factory.annotation.Autowired;
import org.springframework.dao.DuplicateKeyException;
import org.springframework.stereotype.Service;
import org.springframework.transaction.annotation.Transactional;
import com.yhba.service.business.contract.dao.IncontractInfoDao;
```

```
import com.yhba.service.business.notice.NoticeSvc;
@Service
@Transactional(rollbackFor = { Exception.class, SQLException.class })
public class IncontractInfoSvc {
    @Autowired
    IncontractInfoDao incontractInfoDao;
    @Autowired
    NoticeSvc noticeSvc;
    public Object save(Map<String, Object> paramMap) throws SQLException {
            return incontractInfoDao.save(paramMap);
    }
    public List<Map<String, String>> query(Map<String, Object> paramMap) throws
SQLException {
        return incontractInfoDao.query(paramMap);
    }

    public Map<String, Object> queryPagination(Map<String, Object>
paramMap,Map<String, Object> page) throws SQLException {
        return incontractInfoDao.queryPagination(paramMap, page);
    }
    public Object remove(Map<String, Object> paramMap) throws SQLException {
        return incontractInfoDao.remove(paramMap);
    }
    public Object modify(Map<String, Object> paramMap) throws SQLException {
            return incontractInfoDao.modify(paramMap);
    }
    public List<Map<String, String>> queryEnd(Map<String,Object> paramMap) throws
SQLException{
        return incontractInfoDao.queryEnd(paramMap);
    }
}
```

IncontractInfoDao.java:

```
package com.yhba.service.business.contract.dao;
import java.sql.SQLException;
import java.text.SimpleDateFormat;
import java.util.Date;
import java.util.List;
import java.util.Map;
import java.util.HashMap;
import org.apache.commons.lang.RandomStringUtils;
import org.springframework.stereotype.Repository;
import com.yhba.model.persistence.dao.SimpleDaoSupport;
import com.yhba.model.persistence.pagiantion.Pagination;
@Repository
public class IncontractInfoDao extends SimpleDaoSupport {
    public Object save(Map<String, Object> paramMap) throws SQLException {
        paramMap.put("incontractId", getId("incontract_id"));
        //判读是否重复录入
    if(Integer.parseInt(getSqlSession().selectOne("purdetailinfo.check",paramMap).
toString())>0){
            return -1;
        }else{
            return getSqlSession().insert("incontractinfo.insert", paramMap);
        }
    }
```

```
        public List<Map<String, String>> query(Map<String, Object> paramMap) throws
SQLException {
            //fuzzyQuerySupport(paramMap, "employeeCode");
            return getSqlSession().selectList("incontractinfo.queryList", paramMap);
        }
        public Object queryCount(Map<String, Object> paramMap) throws SQLException {
            return getSqlSession().selectOne("incontractinfo.queryEmployCount",
    paramMap);
        }
        public Map<String, Object> queryPagination(Map<String, Object>
paramMap,Map<String, Object> page) throws SQLException {
            Map<String, Object> retMap = new HashMap<String, Object>();
            Pagination pagination = new Pagination(page,queryCount(paramMap));
            paramMap.putAll(pagination.getParameter());
            retMap.put("data", query(paramMap));
            retMap.put("page", pagination.getPagination());
            return retMap;
        }
        public Object remove(Map<String, Object> paramMap) throws SQLException {
            return
Integer.valueOf(getSqlSession().delete("incontractinfo.delete",paramMap));
        }
        public Object modify(Map<String, Object> paramMap) throws SQLException {
            //判读是否重复录入
if(Integer.parseInt(getSqlSession().selectOne("purdetailinfo.check",paramMap).toString
())>0){
                    return -1;
                }else{
                    return
Integer.valueOf(getSqlSession().update("incontractinfo.update",paramMap));
                }
        }
        public List<Map<String, String>> queryEnd(Map<String,Object> paramMap) throws
SQLException{
            Date now = new Date();
            SimpleDateFormat time=new SimpleDateFormat("yyyy-MM-dd");//定义格式
            paramMap.put("nowTime", time.format(now));
             paramMap.put("value", "15");
            return getSqlSession().selectList("incontractinfo.queryEnd", paramMap);
        }
    }
```

IncontractInfoSqlMap.xml：

```
    <?xml version="1.0" encoding="UTF-8"?>
    <!DOCTYPE mapper
        PUBLIC "-//mybatis.org//DTD Mapper 3.0//EN"
        "http://mybatis.org/dtd/mybatis-3-mapper.dtd">
    <mapper namespace="incontractinfo">
        <!-- 记录 -->
        <resultMap id="ResultMap" type="HashMap">
            <result property="incontractId" column="incontract_id" jdbcType="VARCHAR"
javaType="string" />
            <result property="employeeCode" column="employee_code" jdbcType="VARCHAR"
javaType="string" />
            <result property="employeeName" column="employee_name" jdbcType="VARCHAR"
javaType="string" />
```

```
        <result property="securityNo" column="security_no" jdbcType="VARCHAR"
javaType="string" />
        <result property="registerNo" column="register_no" jdbcType="VARCHAR"
javaType="string" />
        <result property="signUp" column="sign_up" jdbcType="VARCHAR"
javaType="string" />
        <result property="signTime" column="sign_time" jdbcType="VARCHAR"
javaType="string" />
        <result property="renewTime" column="renew_time" jdbcType="VARCHAR"
javaType="string" />
        <result property="jobCategory" column="job_category" jdbcType="VARCHAR"
javaType="string" />
        <result property="jobcateName" column="jobcate_name" jdbcType="VARCHAR"
javaType="string" />
        <result property="startTime" column="start_time" jdbcType="VARCHAR"
javaType="string" />
        <result property="stopTime" column="stop_time" jdbcType="VARCHAR"
javaType="string" />
        <result property="operator" column="operator" jdbcType="VARCHAR"
javaType="string" />
        <result property="remark" column="remark" jdbcType="VARCHAR"
javaType="string" />
    </resultMap>
    <sql id="conditionSql">
        <where>
            <if test="employeeCode != null">employee_code = #{employeeCode}</if>
            <if test="employeeName != null"> and employee_name = #{employeeName}</if>
            <if test="securityNo != null"> and security_no = #{securityNo}</if>
            <if test="registerNo != null"> and register_no = #{registerNo}</if>
            <if test="signUp != null"> and sign_up = #{signUp}</if>
            <if test="signTime != null"> and sign_time = #{signTime}</if>
            <if test="renewTime != null"> and renew_time = #{renewTime}</if>
            <if test="jobCategory != null"> and job_category = #{jobCategory}</if>
            <if test="startTime != null"> and start_time = #{startTime}</if>
            <if test="stopTime != null"> and stop_time = #{stopTime}</if>
            <if test="operator != null"> and operator = #{operator}</if>
            <if test="remark != null"> and remark = #{remark}</if>
        </where>
    </sql>
    <!-- 查询记录 -->
    <select id="queryList" parameterType="HashMap" resultMap="ResultMap">
        <include refid="pagination.paginationStart"></include>
        select * from vw_incontract
        <include refid="conditionSql" />
        order by incontract_id DESC
        <include refid="pagination.paginationLast"></include>
    </select>
    <!--查询客户记录数 -->
    <select id="queryEmployCount" parameterType="HashMap" resultType="integer">
        select count(*) from incontract_info
         <include refid="conditionSql" />
    </select>
    <!--查询劳动合同到期 -->
    <select id="queryEnd" parameterType="HashMap" resultMap="ResultMap">
        <include refid="pagination.paginationStart"></include>
        select * from vw_incontract
        where datediff(stop_time,#{nowTime})&lt;= #{value}
```

```
        and datediff(stop_time,#{nowTime})&gt;= 0
        order by incontract_id DESC
        <include refid="pagination.paginationLast"></include>
    </select>
    <!--查询重复记录数 -->
    <select id="check" parameterType="HashMap" resultType="integer">
        select count(*) from incontract_info where incontract_id != #{incontractId}
and
        (employee_code = #{employeeCode} or security_no = #{securityNo})
    </select>
    <!--新增 -->
    <insert id="insert" parameterType="HashMap">
        insert into incontract_info(incontract_id,
        <trim suffixOverrides=",">
            <if test="employeeCode != null">employee_code,</if>
            <if test="employeeName != null">employee_name,</if>
            <if test="securityNo != null">security_no,</if>
            <if test="registerNo != null">register_no,</if>
            <if test="signUp != null">sign_up,</if>
            <if test="signTime != null and signTime != '' ">sign_time,</if>
            <if test="signTime == '' ">sign_time,</if>
            <if test="renewTime != null and renewTime != '' ">renew_time,</if>
            <if test="renewTime == '' ">renew_time,</if>
            <if test="jobCategory != null">job_category,</if>
            <if test="startTime != null and startTime != '' ">start_time,</if>
            <if test="startTime == '' ">start_time,</if>
            <if test="stopTime != null and stopTime != '' ">stop_time,</if>
            <if test="stopTime == '' ">stop_time,</if>
            <if test="operator != null">operator,</if>
            <if test="remark != null">remark</if>
        </trim>
        )values(#{incontractId},
        <trim suffixOverrides=",">
            <if test="employeeCode != null">#{employeeCode},</if>
            <if test="employeeName != null">#{employeeName},</if>
            <if test="securityNo != null"> #{securityNo},</if>
            <if test="registerNo != null">#{registerNo},</if>
            <if test="signUp != null">#{signUp},</if>
            <if test="signTime != null and signTime != '' ">#{signTime},</if>
            <if test="signTime == '' ">null,</if>
            <if test="renewTime != null and renewTime != '' "> #{renewTime},</if>
            <if test="renewTime == '' ">null,</if>
            <if test="jobCategory != null">#{jobCategory},</if>
            <if test="startTime != null and startTime != '' ">#{startTime},</if>
            <if test="startTime == '' ">null,</if>
            <if test="stopTime != null and stopTime != '' ">#{stopTime},</if>
            <if test="stopTime == '' ">null,</if>
            <if test="operator != null">#{operator},</if>
            <if test="remark != null"> #{remark}</if>
        </trim>
        )
    </insert>
    <update id="update" parameterType="HashMap">
        update incontract_info
        <set>
```

```
                <if test="employeeCode != null">employee_code = #{employeeCode},</if>
                <if test="employeeName != null">employee_name = #{employeeName},</if>
                <if test="securityNo != null">security_no = #{securityNo},</if>
                <if test="registerNo != null">register_no = #{registerNo},</if>
                <if test="signUp != null">sign_up = #{signUp},</if>
                <if test="signTime != null and signTime != '' ">sign_time =
#{signTime},</if>
                <if test="signTime == '' ">sign_time = null,</if>
                <if test="renewTime != null and renewTime != '' ">renew_time =
#{renewTime},</if>
                <if test="renewTime == '' ">renew_time = null,</if>
                <if test="jobCategory != null">job_category = #{jobCategory},</if>
                <if test="startTime != null and startTime != '' ">start_time =
#{startTime},</if>
                <if test="startTime == '' ">start_time = null,</if>
                <if test="stopTime != null and stopTime != '' ">stop_time =
#{stopTime},</if>
                <if test="stopTime == '' ">stop_time = null,</if>
                <if test="operator != null">operator = #{operator},</if>
                <if test="remark != null">remark = #{remark}</if>
            </set>
            where incontract_id = #{incontractId}
        </update>
        <!-- 删除 -->
        <delete id="delete" parameterType="HashMap">
            delete from incontract_info where incontract_id in (${incontractIds})
        </delete>
    </mapper>
```

合同解除模块与合同签订模块的内容逻辑大致相同，这里不做详细描述，下面我们给出劳动合同管理模块的效果图。

合同签订模块主窗口如图 9.15 所示。

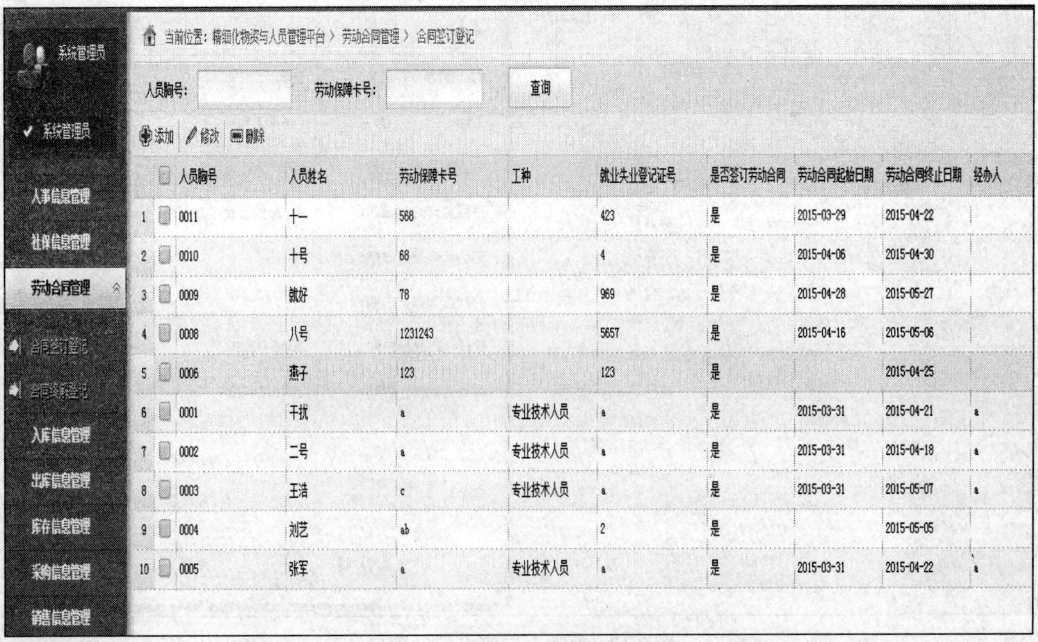

图 9.15　合同签订模块主窗口

添加合同签订记录窗口如图 9.16 所示。

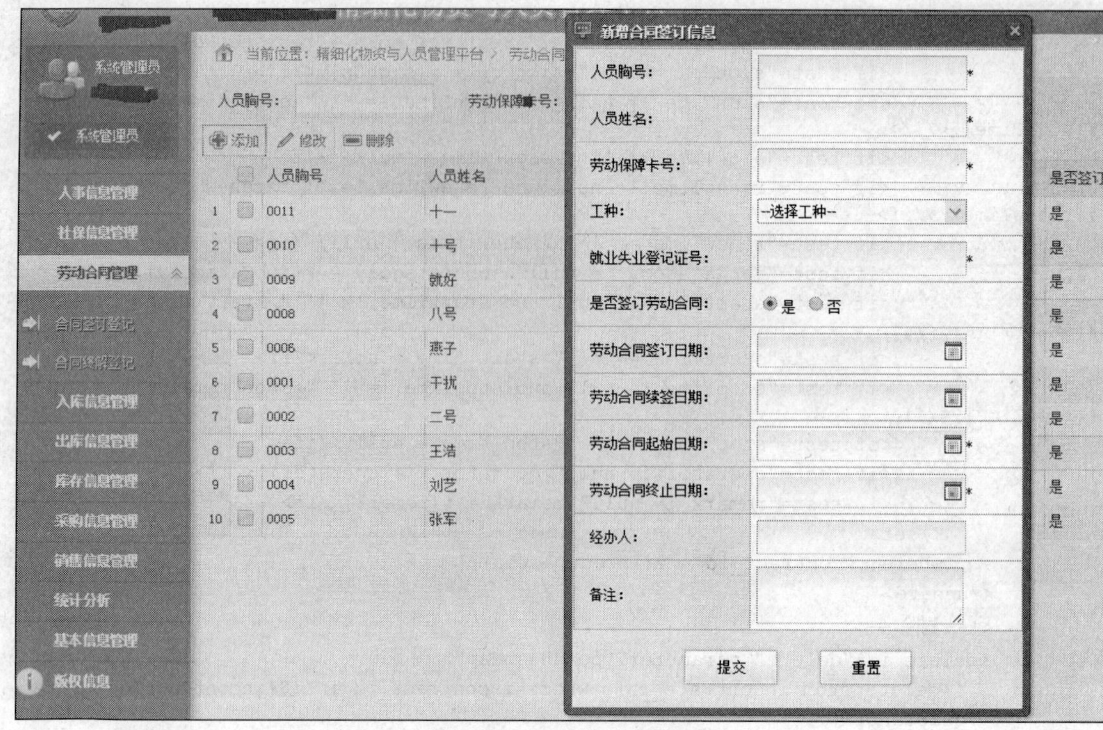

图 9.16　添加合同签订记录窗口

修改合同签订记录如图 9.17 所示。

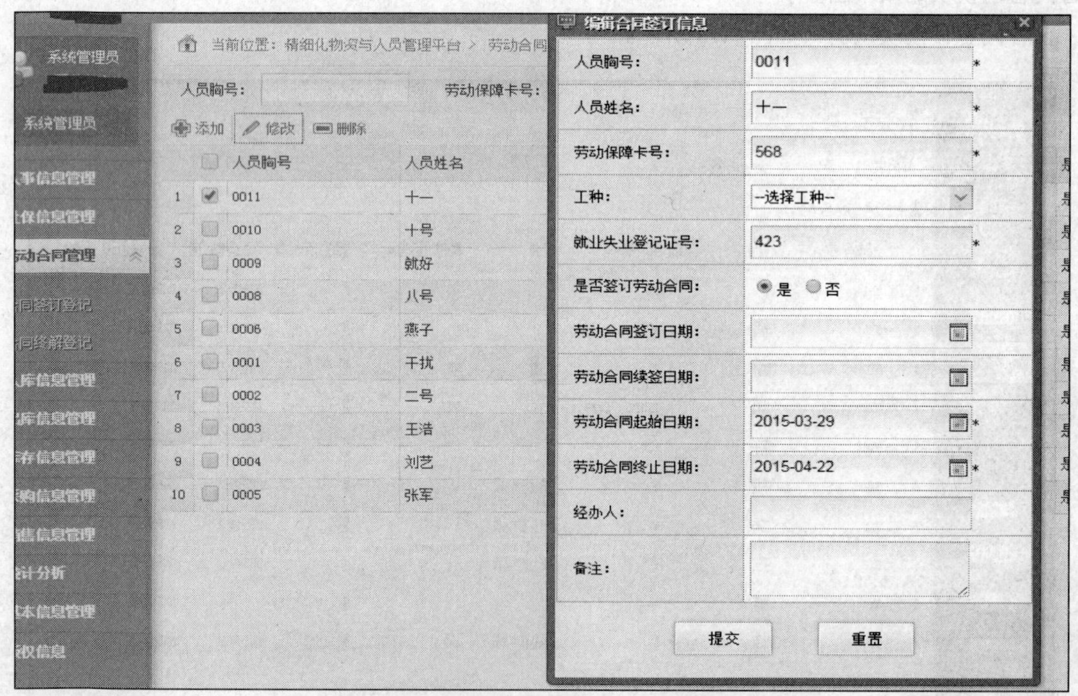

图 9.17　修改合同签订记录窗口

删除合同签订记录如图 9.18 所示。

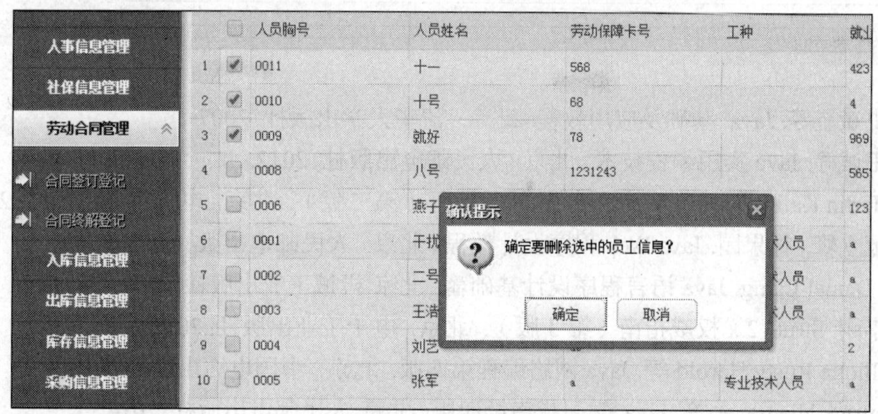

图 9.18　删除合同签订记录窗口

查看合同签订信息如图 9.19 所示。

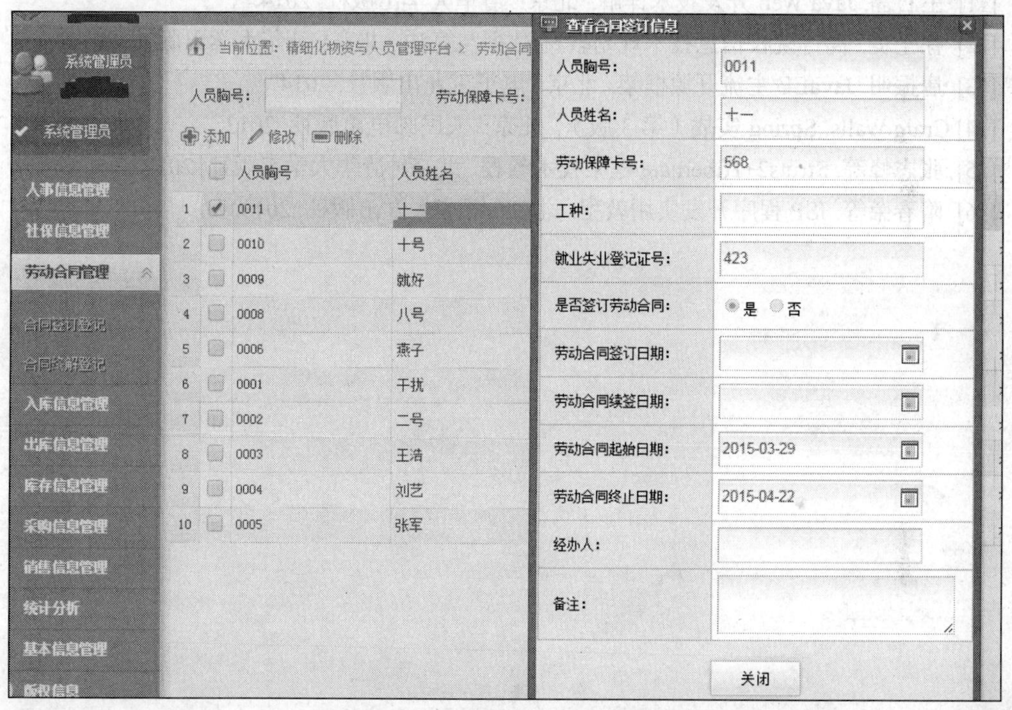

图 9.19　查看合同签订信息窗口

参考文献

[1] 史斌星等. Java 基础及应用教程. 北京: 清华大学出版社, 2007.

[2] 王诚等. Java 实用编程技术. 北京: 人民邮电出版社, 2012.

[3] Martin Kalin 等. Java Web 服务:构建与运行（第二版）. 北京: 电子工业出版社, 2014.

[4] 范立峰, 林果园. Java Web 程序设计教程. 北京: 人民邮电出版社, 2010.

[5] Y.Daniel Liang. Java 语言程序设计基础篇. 北京:机械工业出版社, 2011.

[6] 李刚. Struts 2.x 权威指南（第 3 版）. 北京: 电子工业出版社, 2012.

[7] Elliotte Rusty Harold 等. Java 网络编程第四版. 北京: 中国电力出版社, 2014.

[8] Patrick Niemeyer 等. Java 学习指南第四版. 北京:人民邮电出版社, 2014.

[9] 孙卫琴, 李洪成. Tomcat 与 Java Web 开发技术详解. 北京:电子工业出版社, 2006.

[10] 张志峰等. Java Web 技术整合应用与项目实战. 北京: 清华大学出版社, 2013.

[11] 王石磊. Java Web 开发技术详解. 北京: 清华大学出版社, 2014.

[12] 徐士良. 高等院校信息技术规划教材:数据与算法. 北京: 清华大学出版社, 2014.

[13] 唐振明. JavaEE 主流开源框架. 北京: 电子工业出版社, 2014.

[14] Craig Walls. Spring 实战（第 3 版）. 北京: 人民邮电出版社, 2013.

[15] 张志峰等. Struts2+Hibernate 框架技术教程. 北京: 清华大学出版社, 2012.

[16] 隋春荣等. JSP 程序开发实用教程. 北京: 清华大学出版社, 2013.